METHODS IN MOLECULAR BIOLOGY

For further volumes:
http://www.springer.com/series/7651

Clinical Applications of PCR

Third Edition

Edited by

Rajyalakshmi Luthra, Rajesh R. Singh, and Keyur P. Patel

Department of Hematopathology, The University of Texas MD Anderson Cancer Center, Houston, TX, USA

Editors
Rajyalakshmi Luthra
Department of Hematopathology
The University of Texas MD Anderson Cancer Center
Houston, TX, USA

Rajesh R. Singh
Department of Hematopathology
The University of Texas MD Anderson Cancer Center
Houston, TX, USA

Keyur P. Patel
Department of Hematopathology
The University of Texas MD Anderson Cancer Center
Houston, TX, USA

ISSN 1064-3745 ISSN 1940-6029 (electronic)
Methods in Molecular Biology
ISBN 978-1-4939-3358-7 ISBN 978-1-4939-3360-0 (eBook)
DOI 10.1007/978-1-4939-3360-0

Library of Congress Control Number: 2015957392

Springer New York Heidelberg Dordrecht London

Printed on acid-free paper

Humana Press is a brand of Springer
Springer Science+Business Media LLC New York is part of Springer Science+Business Media (www.springer.com)

Preface

Amplification of nucleic acids using polymerase chain reaction (PCR) is the backbone of most techniques used in genome analysis. Since its invention in 1983, the improvements, variations, and applications of PCR have grown exponentially making it an indispensable technique in molecular biology. More recently, the application of PCR has moved from amplifying a single region using a pair of primers flanking the areas of interest to medium and high-multiplexed approaches where hundreds to thousands of primers amplify multiple areas of interest in a single reaction. This capability has really provided impetus towards an expanded scope of applications for genomic sequencing. With the onset of massively parallel sequencing capabilities of Next Generation Sequencing (NGS), PCR is again proving to be the tool of choice in every step of NGS, be it target enrichment or clonal amplification prior to sequencing. Consequently, the application of PCR in the study of genomics and transcriptomics has tremendous impact, not only for discovery but also for routine clinical applications, and forms the cornerstone of personalized medicine.

In the third edition of this book, titled *Clinical Applications of PCR*, we have tried to highlight a wide variety of clinical applications of PCR such as detecting DNA methylation, detection of viruses and protozoa in infectious diseases, estimation of gene copy number aberrations, primer extension coupled with mass spectroscopy, and high-throughput NGS techniques. We like to thank all contributing authors of the book chapters and hope the readers find this collection of topics and the detailed methodology provided useful in understanding the principle behind each application and for implementation in the laboratory.

Houston, TX, USA

Rajyalakshmi Luthra
Rajesh R. Singh
Keyur P. Patel

Contents

Contributors

KENNETH D. ALDAPE • *Princess Margaret Cancer Center and Ontario Cancer Institute, Toronto, ON, Canada*

MARIE E. ARCILA • *Department of Pathology, Memorial Sloan Kettering Cancer Center, New York, NY, USA*

HUI CHEN • *Department of Pathology, The University of Texas MD Anderson Cancer Center, Houston, TX, USA*

JUAN LUIS CONCEPCIÓN • *Laboratorio de Enzimología de Parásitos, Universidad de Los Andes, Mérida, Venezuela*

EGLYS GONZÁLEZ-MARCANO • *Laboratorio de Enzimología de Parásitos, Universidad de Los Andes, Mérida, Venezuela; Fundación Jacinto Convit, Caracas, Venezuela*

RASHMI S. GOSWAMI • *Department of Laboratory Hematology, University Health Network, Toronto General Hospital, Toronto, ON, Canada*

MEERA HAMEED • *Department of Pathology, Memorial Sloan Kettering Cancer Center, New York, NY, USA*

KAUSAR J. JABBAR • *Department of Hematopathology, The University of Texas MD Anderson Cancer Center, Houston, TX, USA*

RASHMI KANAGAL-SHAMANNA • *Department of Hematopathology, The University of Texas M.D. Anderson Cancer Center, Houston, TX, USA*

HIROTOMO KATO • *Laboratory of Parasitology, Department of Disease Control, Graduate School of Veterinary Medicine, Hokkaido University, Sapporo, Japan*

SANAM LOGHAVI • *Department of Hematopathology, The University of Texas MD Anderson Cancer Center, Houston, TX, USA*

MARÍA ELIZABETH MÁRQUEZ • *Laboratorio de Enzimología de Parásitos, Universidad de Los Andes, Mérida, Venezuela*

MEENAKSHI MEHROTRA • *Department of Hematopathology, The University of Texas MD Anderson Cancer Center, Houston, TX, USA*

ALBERTO PANIZ MONDOLFI • *Microbiology Laboratory, Department of Laboratory Medicine, Yale New Haven Hospital, Yale University School of Medicine, New Haven, CT, USA; Department of Hematopathology, The University of Texas MD Anderson Cancer Center, Houston, TX, USA*

KHEDOUDJA NAFA • *Department of Pathology, Memorial Sloan Kettering Cancer Center, New York, NY, USA*

CHI YOUNG OK • *Department of Hematopathology, The University of Texas MD Anderson Cancer Center, Houston, TX, USA*

KEYUR P. PATEL • *Department of Hematopathology, The University of Texas MD Anderson Cancer Center, Houston, TX, USA*

SINCHITA ROY-CHOWDHURI • *Department of Pathology, Cytopathology and Molecular Pathology, MD Anderson Cancer Center, The University of Texas, Houston, TX, USA*

ALAA A. SALIM • *Department of Hematopathology, The University of Texas MD Anderson Cancer Center, Houston, TX, USA*

CHARANJEET SINGH • *Department of Pathology, The University of Texas MD Anderson Cancer Center, Houston, TX, USA*
RAJESH R. SINGH • *Department of Hematopathology, The University of Texas MD Anderson Cancer Center, Houston, TX, USA*
ALAIN R. THIERRY • *IRCM, Institut de Recherche en Cancérologie de Montpellier, Montpellier, France; INSERM, U1194, Montpellier, France; Université de Montpellier, Montpellier, France; Institut Régional du Cancer de Montpellier, Montpellier, France; DiaDx SARL, Montpellier, France*
KHALIDA WANI • *Department of Translational Molecular Pathology, The University of Texas MD Anderson Cancer Center, Houston, TX, USA*
ZHUANG ZUO • *Department of Hematopathology, The University of Texas MD Anderson Cancer Center, Houston, TX, USA*

Chapter 1

A Targeted Q-PCR-Based Method for Point Mutation Testing by Analyzing Circulating DNA for Cancer Management Care

Alain R. Thierry

Abstract

Circulating cell-free DNA (cfDNA) is a valuable source of tumor material available with a simple blood sampling enabling a noninvasive quantitative and qualitative analysis of the tumor genome. cfDNA is released by tumor cells and exhibits the genetic and epigenetic alterations of the tumor of origin. Circulating cell-free DNA (cfDNA) analysis constitutes a hopeful approach to provide a noninvasive tumor molecular test for cancer patients. Based upon basic research on the origin and structure of cfDNA, new information on circulating cell-free DNA (cfDNA) structure, and specific determination of cfDNA fragmentation and size, we revisited Q-PCR-based method and recently developed a the allele-specific-Q-PCR-based method with blocker (termed as Intplex) which is the first multiplexed test for cfDNA. This technique, named Intplex® and based on a refined Q-PCR method, derived from critical observations made on the specific structure and size of cfDNA. It enables the simultaneous determination of five parameters: the cfDNA total concentration, the presence of a previously known point mutation, the mutant (tumor) cfDNA concentration (ctDNA), the proportion of mutant cfDNA, and the cfDNA fragmentation index. Intplex® has enabled the first clinical validation of ctDNA analysis in oncology by detecting KRAS and BRAF point mutations in mCRC patients and has demonstrated that a blood test could replace tumor section analysis for the detection of KRAS and BRAF mutations. The Intplex® test can be adapted to all mutations, genes, or cancers and enables rapid, highly sensitive, cost-effective, and repetitive analysis. As regards to the determination of mutations on cfDNA Intplex® is limited to the mutational status of known hotspot mutation; it is a "targeted approach." However, it offers the opportunity in detecting quantitatively and dynamically mutation and could constitute a noninvasive attractive tool potentially allowing diagnosis, prognosis, theranostics, therapeutic monitoring, and follow-up of cancer patients expanding the scope of personalized cancer medicine.

Key words Circulating DNA, Cell-free DNA, Plasma, Mutation, Q-PCR, Cancer, KRAS, BRAF, Sensitivity, Diagnostic

1 Introduction

1.1 Potential of cfDNA Analysis in Oncology

The discovery of circulating cell-free DNA (cfDNA) in the human circulatory system has led to intensive research on its use in various clinical fields. The utmost development of cfDNA consists in the recent approval of embryo sex testing. CfDNA was discovered in 1948 by Mandel and Metais [1] although at the time it did not

Rajyalakshmi Luthra et al. (eds.), *Clinical Applications of PCR*, Methods in Molecular Biology, vol. 1392,
DOI 10.1007/978-1-4939-3360-0_1, © Springer Science+Business Media New York 2016

attract much curiosity. However, 30–40 years later, the interest of cfDNA was demonstrated by several groups including Leon et al. [2], who found that cfDNA concentration significantly increased in cancer subjects, and Stroun et al. [3], who described a proportion of cfDNA that was tumor derived and carried its molecular characteristics, leading to the concept of a "liquid biopsy." Therefore, cfDNA analysis could provide diagnostic, prognostic, and theragnostic information [4–6]. Several researchers are intensively developing techniques that allow detection and characterization of genetic and epigenetic alterations of tumor cells using cfDNA analysis in the plasma or serum of cancer subjects. Such techniques could revolutionize the management of cancer-patient care through the detection of mutations leading to resistance to targeted therapies, personalized therapeutic monitoring, and noninvasive follow-up of the disease. CfDNA analysis is currently used in prenatal diagnosis [4] and shows potential for analysis in other clinical fields, such as autoimmune diseases, trauma, sepsis, or myocardial infarction [5].

Despite intensive research, few cfDNA-based tests have been translated to clinical practice. Several techniques are under development to detect and characterize cfDNA in cancer subjects including restriction fragment length polymorphism, direct sequencing, high-resolution melting analysis, digital PCR, cold PCR, and other techniques usually used for tumor tissue analysis. Nevertheless, cfDNA concentration has not yet been validated as a cancer biomarker because the literature reveals conflicting data: plasma cfDNA concentrations in cancer subjects range from a few ng/ml to several thousand ng/ml, which overlap with the concentration range for healthy individuals [6]. Furthermore, estimation of cfDNA fragmentation in cancer subjects has been found to be lower [7], equivalent [8], or higher [9–11] than in control subjects. These discrepancies may be explained by the lack of fundamental knowledge of cfDNA. Indeed, the cognitive aspects of cfDNA are still not identified and elucidated: the respective contributions of different potential release mechanisms of cfDNA (apoptosis, necrosis, phagocytosis, extracellular DNA traps, active release, etc.) and the cfDNA structures (part of chromatin, nucleosomes, nucleoprotein complex, exosomes, apoptotic bodies, etc.) have not been clearly defined [5–11].

Total circulating DNA concentration was envisaged for long time as a potential cancer biomarker, but cfDNA concentration values from healthy and cancer individuals have been shown to partly overlap, precluding its development for clinical use [6, 12]. Recently using current methods and a specific PCR system design, statistically significant differences in cfDNA have been shown between healthy subjects and cancer subjects [13]. CfDNA analysis constitutes a hopeful approach to provide a noninvasive tumor molecular test for cancer patients, especially for detecting genetic or epigenetic alterations, such as point mutations or

Table 1
Tumor tissue analysis vs. plasma DNA analysis in regard to management care of cancer patient

Tumor tissue analysis	Plasma DNA analysis
Con: – Sampling bias due to: Intra-tumor heterogeneity Inter-tumor heterogeneity (Same for meta- or synchronous tumor) – Not always available – Sometimes long data turnaround time – Biopsy difficult to assess in some cancer and not repeatable except risky biopsy	Con: – Cannot distinguish: Synchronous metastasis and primary between metastasis of different sites
Pro: – Can be associated with anatomo-pathological observation – Is directly associated to a specific tumor tissue (i.e., in case of various metastatic sites)	Pro: – Instant picture of the tumor – Quantitative multimarker analysis – Enables longitudinal analysis for: Emergence of mutations Surveillance of the recurrence Minimal residual disease Treatment monitoring – Prognostics – Determination of the sensitivity threshold of the targeted therapies

single-nucleotide polymorphisms, microsatellite alterations, methylations, or copy number variations [12]. In particular, detecting point mutations in numerous cancer types has been investigated intensely during the last decade [14–16].

As determination of some mutational status such as of ALK, EGFR, BRAF, or KRAS was found as a prerequisite for selecting to specific targeted therapies patients with lung, melanoma, or colorectal cancer, it was the subject of the first application of cfDNA. Mutational status testing is conventionally performed from tumor tissue, principally from primary tumors and to lesser extent from biopsy when available. As presented in Table 1 plasma DNA analysis shows numerous advantages over tumor tissue analysis in regard to management care of cancer patient.

Detection of somatic point mutations must be very sensitive, particularly in the case of ctDNA: it has been postulated that a tiny fraction down to 0.001 % of the ctDNA of a cancer patient (down to a few dozen of copies/ml plasma) carries the mutation [10]. For this reason, numerous modifications of the Q-PCR technology have been described, such as ARMS-PCR, TaqMAMA, and FLAG-PCR. These technologies require the use of modified bases, specific enzymes, or additional procedures in addition to the reagents. Other advanced technologies have emerged, such as parallel sequencing [9, 10] or the beaming technology, which need expertise and specific, sophisticated equipment [10, 12].

1.2 Background and Pioneering Observations

Our group has worked since 2005 on circulating DNA and has provided pioneering observations on the structure and origins of tumor-derived cfDNA [10, 11, 17]. We investigated in depth the CtDNA size distribution in order to determine the value of ctDNA as a potential theranostic tool. We used a highly specific Q-PCR assay and athymic nude mice xenografted with human colon cancer cells and compared our results with those obtained from the plasma of metastatic CRC patients [10]. Fragmentation and concentration of tumor-derived ctDNA were positively correlated with tumor weight. CtDNA quantification by Q-PCR depended on the amplified target length and was optimal for 60–100 bp fragments [11, 18]. Q-PCR analysis of plasma samples from xenografted mice and cancer patients showed that tumor-derived ctDNA exhibited a specific amount profile based on ctDNA size and significantly higher ctDNA fragmentation [11]. Metastatic colorectal patients ($n = 12$) showed nearly fivefold higher mean ctDNA fragmentation than healthy individuals ($n = 16$). In addition, we demonstrated that nucleic fragment size is crucially important when analyzing cell-free nucleic acid and, in particular, when determining its concentration. Specific single-step detection of mutation by measuring the ctDNA concentration is possible by comparing the concentration determined by targeting a short mutated fragment (<100 bp) and a non-mutated fragment of similar size (±10 %) [11, 17].

In order to provide a quick and accurate genotyping, we revisited the Q-PCR technique swimming against the tide of prevailing approaches involving high-tech methods in designing the Intplex® method. Intplex® is an allele-specific Q-PCR system with an oligo-blocker using an ingenious primer design with characteristics based on previously made structural observations. Intplex is the first multiplexed test for cfDNA. It determines simultaneously five parameters in a single run: (1) total cfDNA concentration, (2) mutant (tumor) cfDNA concentration, (3) proportion of mutant cfDNA, (4) cfDNA fragmentation index, and (5) detection of previously known point mutations. Here, we described in detail the technical aspects of the later parameter.

2 Materials

1. PCR primers (Table 2).
2. Super Mix SYBR Green (Bio-Rad, Hercules, CA, USA).
3. CFX manager software (Bio-Rad).
4. Human placenta DNA (Sigma Aldrich, Saint Louis, Mo, USA).
5. Qubit Fluorimeter (Invitrogen, Carlsbad, CA, USA).

3 Methods

3.1 Intplex® Design Strategy

CfDNA quantification by Q-PCR critically depends on the amplified target length and was optimal for 60–100 bp fragments [11]. Q-PCR analysis of plasma samples from xenografted mice [10] and cancer patients [11] showed that tumor-derived cfDNA exhibited a specific amount profile based on cfDNA size and significantly higher cfDNA fragmentation than healthy individuals. Based on the abovementioned observations on cfDNA size, we developed a novel Q-PCR-based method. The test (Intplex®), a single-step allele-specific Q-PCR with blocker method that involves a specifically designed primer system, is combined with our novel technique for point-mutation detection. Firstly, the test involves the amplification at the extremities of a ~300 bp region of two shorter same-size regions (in the range of 60–100 bp, (±10 %), one corresponding to a mutated sequence and the other to a wild-type sequence (Fig. 1). Secondly, the test discriminates highly and specifically between mutant and wild-type alleles using a blocking 3′-phosphate-modified oligonucleotide and low *Tm* primer [17, 19] (Table 2).

Detection of somatic point mutations must be very sensitive, particularly in the case of ctDNA: it has been postulated that a tiny fraction (only 0.01–10 %) of the ctDNA of a cancer patient carries the mutation [5, 12, 14]. For this reason, numerous modifications of the Q-PCR technology have been described, such as ARMS-PCR, TaqMAMA, and FLAG-PCR [13]. These technologies require the use of modified bases, specific enzymes, or additional procedures in addition to the reagents. Other advanced technologies have emerged, such as parallel sequencing [9, 10] or the beaming technology, which need expertise and specific, sophisticated equipment [14]. Note that COLD-PCR is really an efficient and smart method to enrich mutant fragment amount to be detected and very high sensitivity is obtained, and can be used with our primer systems. However, this method lacks the possibility of being quantitative impeding the determination of mutation allele frequency, or the concentration of mutant cfDNA.

3.2 Blood Collection and cfDNA Extraction

Preanalytical operating procedures followed previously reported guidelines [19].

3.3 Intplex® Primer Design

The sequences and the characteristics of the selected primers [11] are presented in Table 2. Primers were designed on Primer3 software and subsequently selected according to local-alignment analyses with BLAST and mfold program. Oligonucleotides were synthesized and purified by HPLC by Eurofins (Ebersberg,

Table 2
Characteristics of the selected primers and of the amplicons obtained [18]

Species	Gene	Primer name	Direction	Sequence 5′-3′	*Tm* (°C)	Amplicon size (bp)
Human	*BRAF*	Braf A1 conv k	Sense	TTATTGACTCTAAGAGGAAAGATGAA	56.9	105
		Braf A2 conv k	Antisense	GAGCAAGCATTATGAAGAGTTTAGG	59.7	–
		Braf V600E conv k	Sense	GATTTTGGTCTAGCTACAGA	53.2	97
		Braf B2 conv k	Antisense	TAGCCTCAATTCTTACCATCCACA	59.3	–
		Braf blocker	Sense	GCTACAGTGAAATCTCGATGG—PHO		
Human	*KRAS*	Kras A1 inv k	Sense	GCCTGCTGAAAATGACTGA	54.5	61
		Kras G12D Inv k	Antisense	CTCTTGCCTACGCCAT	51.7	–
		Kras G12A Inv k	Antisense	CTCTTGCCTACGCCAG	54.3	–
		Kras G12S Inv k	Antisense	TCTTGCCTACGCCACT	51.7	60
		Kras G12C Inv k	Antisense	TCTTGCCTACGCCACA	51.7	–
		Kras G13D Inv k	Antisense	GCACTCTTGCCTACGT	51.7	64
		Kras G12V Inv k	Antisense	CTCTTGCCTACGCCAA	51.7	61
		Kras G12R Inv k	Antisense	TCTTGCCTACGCCACG	54.3	62
		Kras B1 inv k	Sense	CCTTGGGTTTCAAGTTATATG	54.0	67
		Kras B2 inv k	Antisense	CCCTGACATACTCCCAAGGA	59.4	–
		Kras blocker	Antisense	GCCTACGCCACCAGCTC—PHO		

PHO (phosphate) modification

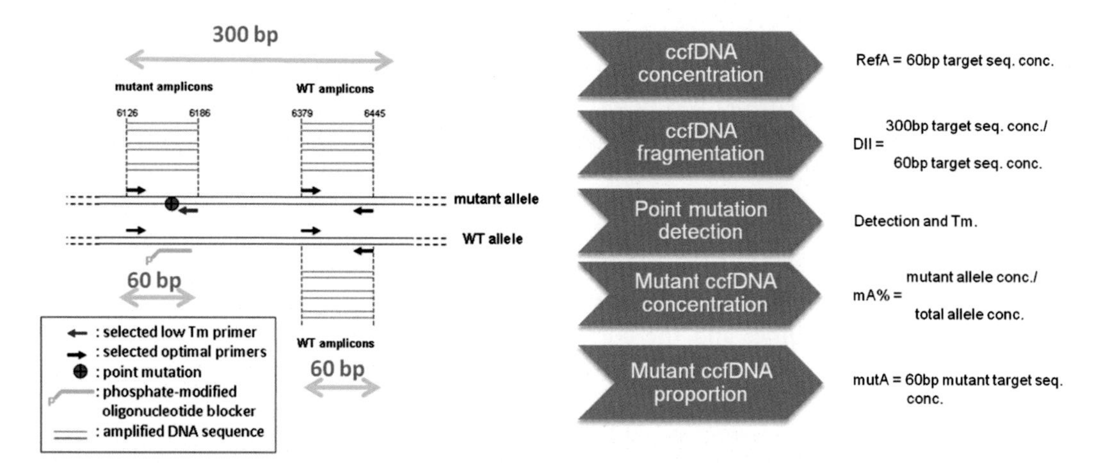

Fig. 1 Primer set design analysis of the codon 12 KRAS mutation testing and diagnostic analytical parameters of the Intplex method

Germany). Quality control of the oligonucleotides was performed by MALDI-TOF.

3.4 cfDNA Quantification by Quantitative PCR

The Q-PCR assays were performed according to the MIQE guidelines [20]. Q-PCR amplifications were carried out at least in duplicate in a reaction volume of 25 μl on a CFX96 instrument using the CFX manager software (Bio-Rad). Each PCR reaction mixture was composed of 12.5 μl PCR mix (Bio-Rad Super mix SYBR Green), 2.5 ml of each amplification primer (3 pmol/ml, final concentration), 2.5 μl PCR-analyzed water, and 5 μl DNA extract. Thermal cycling consisted of three repeated steps: a 3-min hot-start polymerase activation-denaturation step at 95 °C, followed by 40 repeated cycles at 95 °C for 10 s, and then at 60 °C for 30 s. Melting curves were obtained by increasing the temperature from 55 to 90 °C with a plate reading every 0.2 °C. The concentration was calculated from Cq detected by Q-PCR and also a control standard curve on DNA of known concentration and copy number (Sigma-Aldrich). Serial dilutions of genomic DNA from human placenta cells (Sigma) were used as a standard for quantification and their concentration and quality were assessed using a Qubit spectrofluorimeter (Invitrogen). All the samples were analyzed in duplicate, and each assay was repeated at least once. The cfDNA concentrations were normalized to the precise concentration of a genomic DNA sample amplified using the same primer set in both cases. The coefficient of variation, calculated from two experiments ($n = 12$), was 24 % for the cfDNA extraction and the qPCR analysis.

3.5 Mutation Analysis

The concentration obtained when targeting the mutated sequence corresponded to the concentration of the alleles bearing the mutation (mA). The concentration obtained when targeting the wild-type (WT) sequence located at 300 bp from the position of the point mutation corresponded to the total cfDNA (WT plus mutated cfDNA), denoted refA. The proportion of mutant allele (mA%) was determined by quantifying the relative ratios between mA and refA (Fig. 1). Each Q-PCR experiment was carried out after validating the internal technical controls. Non-template controls for each primer set targeting KRAS or BRAF mutations were included in every experimental run in this study. Positive-control DNA was extracted from cell lines bearing the targeted KRAS and BRAF mutations. The respective correspondence between cell lines and the corresponding mutation was further checked. Note that Intplex is at this time the only method for testing point mutation which has been clinically validated in a large cohort (>100 patients in a blinded prospective multicenter study) and under stringent diagnostic criteria (STARD).

4 Notes

4.1 Performance and Unmet Sensitivity

Because mutant ctDNA can be present at very low frequency (down to 0.01 % of total cfDNA) sensitivity is a critical issue when analyzing cfDNA. Mutant allele concentration value as determined by Intplex was reliably determined up to a dilution of one mutated copy in 20,000 copies of WT DNA (0.01 %) as observed with using genomic DNA bearing *BRAF* V600E (Table 3 and Fig. 2). Note that 0.01 % corresponds to the amplification of one copy in the assay. These selectivity and detection limits are superior to technologies such as ARMS combined with the Scorpions Q-PCR or dye terminator sequencing [6] and are equivalent to the beaming technology or digital PCR [14, 21, 22]. Percentage of non-specificity was calculated as less than 0.01 % for *BRAF* V600E and *KRAS* mutant primer sets [23] (Figs. 2 and 3). Intplex can be adapted to all point mutations of interest.

Intplex® enables single-copy detection of variant alleles down to an unprecedented sensitivity of ≥0.005 % mutant-to-WT ratio which is slightly higher than that of digital PCR or beaming methods and much higher than sequencing or Next Generation Sequencing (0.5 %). It allows the detection of a single-cfDNA fragment under Poisson law. The detection limit of this method is two copies/ml plasma surpassing by more than 100-fold the sensitivity of the available next-generation sequencing methods [4, 15]. Although NGS is being proposed as technical solution for detecting point mutation in ctDNA, its sensitivity is limited by the inherent error rate of the sequencer, as incorrectly read bases might be mistaken for true mutant copies, and consequently Q-PCR provides a sensitivity capability that cannot be matched.

Intplex® success rate is 100 % in more than 300 stage I–IV CRC and 100 healthy individual plasma samples. The mean

Table 3

Intplex® technical characteristics for the detection of KRAS codon 13 and 13 and BRAF V600E mutations

Preclinical requirements:
Plasma volume: 4 ml
Plasma shipped in frozen state
Information provided:
Presence of a specific mutation: KRAS G12V, G12R, G12D, G12C, G12S, G12A, G13D
: BRAF V600E (if requested)
Mutant cfDNA concentration
Relative tumor mutation load
Total (WT) cfDNA concentration:
Fragmentation index: cfDNA extract quality control
Quality controls
cfDNA extract quality control by the level of the total cfDNA concentration (>2 ng/ml plasma) and by the fragmentation index
PCR analytical control by comparing total cfDNA concentration when targeting one WT sequence of two different genes (BRAF and KRAS)
PCR analytical control by positive and negative internal quality controls for each mutation
PCR analytical control by using a standard curves with six points at each run
PCR amplification is always confirmed by the specific melting temperature of the internal positive control (±0.2 °C)
Analytical parameters:
Sensitivity (mutant/WT copy number): KRAS G12V (0.005 %)
G12R (0.01 %)
G12C (0.01 %)
G12D (0.005 %)
G12S (0.01 %)
G12A (0.01 %)
G13D (0.008 %)
BRAF V600E (0.004 %)
Allele frequency range on cancer patient plasma sample: 0.0013–0.2 %
Quantitative detection down to 0.05 ng/ml plasma Qualitative detection down to 0.02 ng/ml plasma
Specificity: 100 % as tested on WT and mutant cell lines and WT healthy individuals (n = 50)
Reproducibility: 100 % as tested by two or three different manipulators (n = 31)
Mean data turnaround after plasma sample reception: 2 days

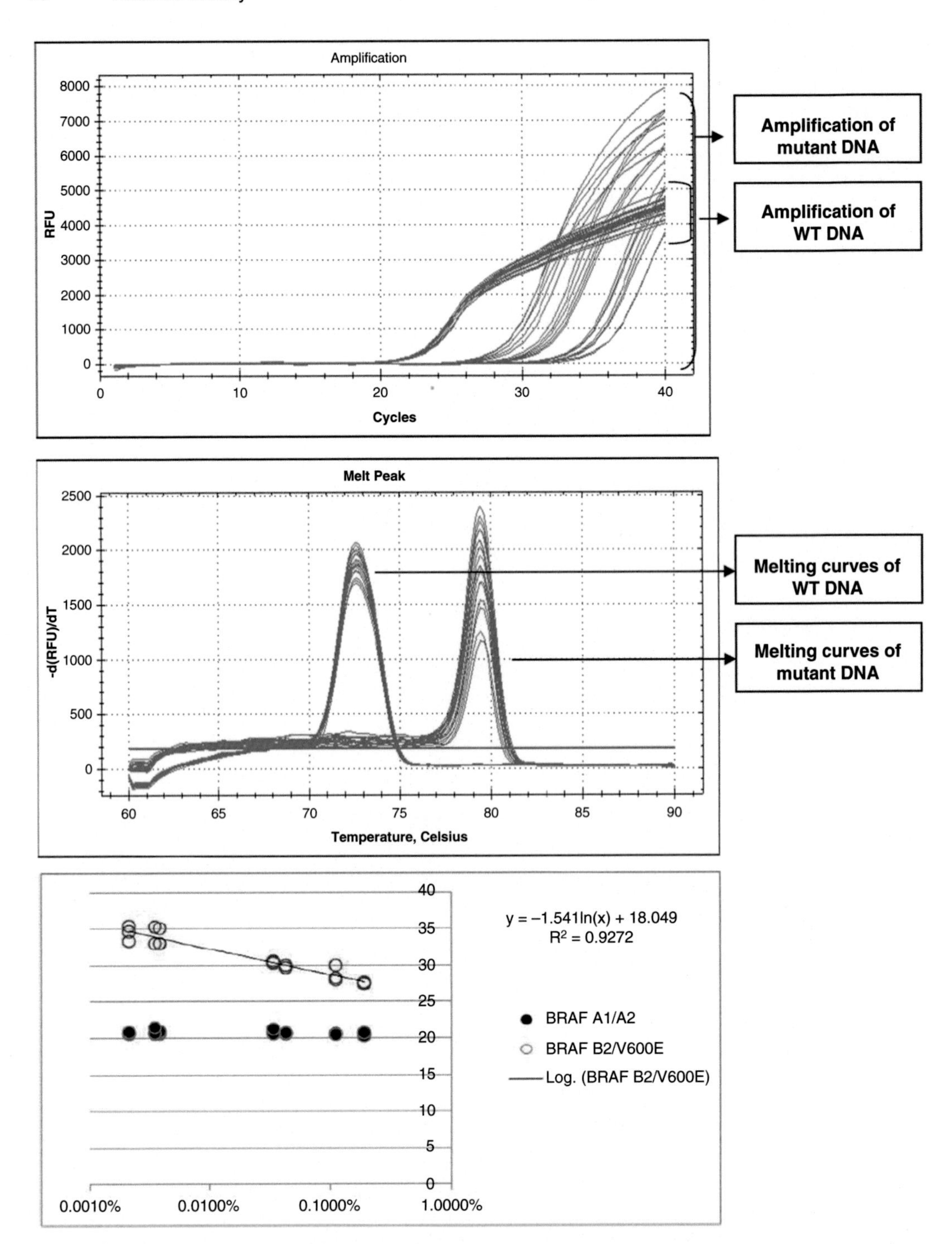

Fig. 2 Sensitivity of the Intplex test for BRAF V600E as determined by analyzing mutant/WT copy number. Sensitivity limit of 0.0038 % (0.92 mutant copies in 25,946 copies of WT DNA ADN)

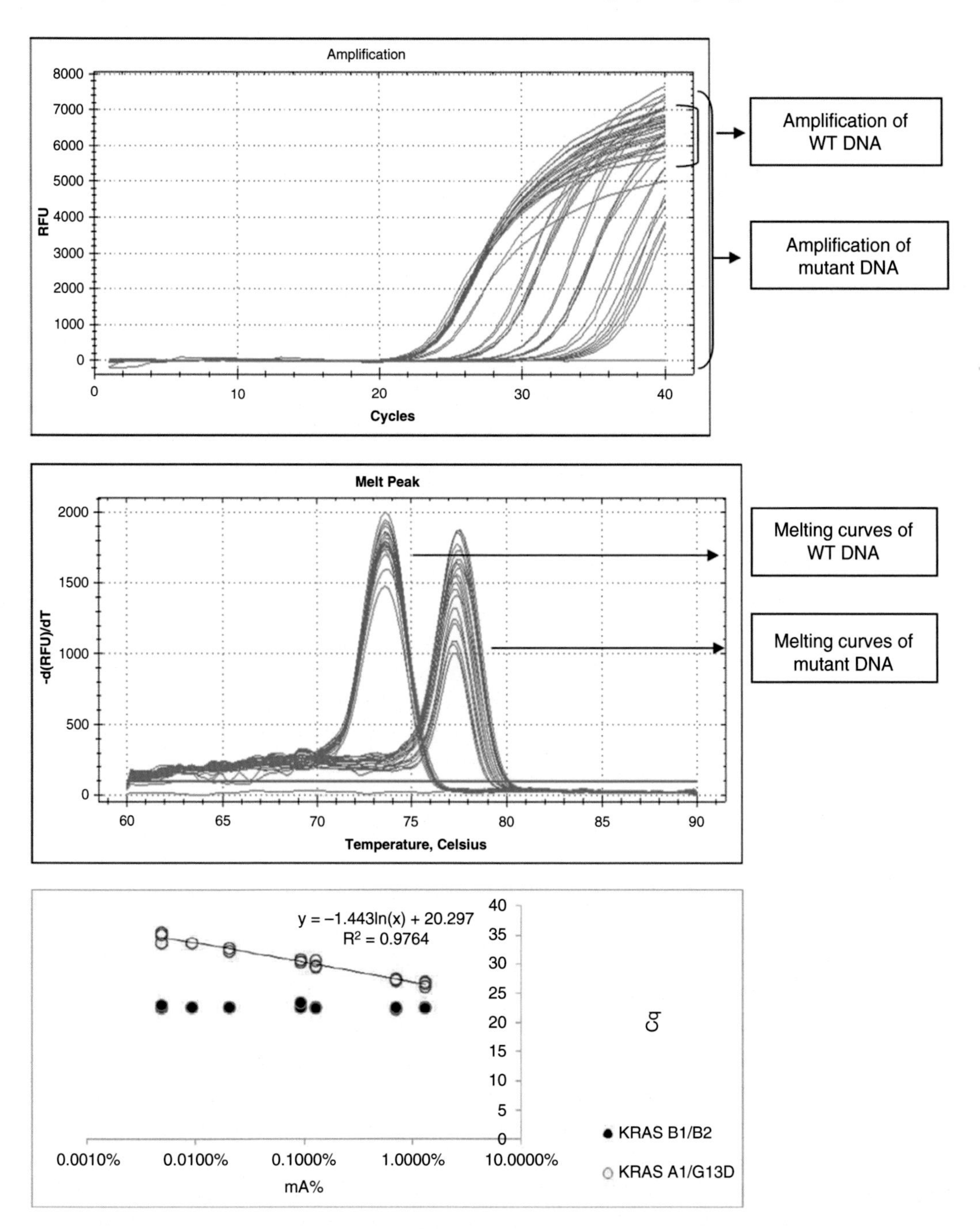

Fig. 3 Sensitivity of the Intplex test for KRAS G13D as determined by analyzing mutant/WT copy number. Sensitivity limit of 0.005 % (0.96 mutant copies in 19,741 copies of WT DNA ADN)

turnaround time is 1 day for codon 12 and 13 KRAS and BRAF V600E mutations and 2 days for the multiplex analysis.

Quality control is one of the assets of the used method. First, a stringent pre-analytical guideline ensures integrity of plasma DNA and potentially excludes samples because of non-proper blood or plasma handling. Quality control for the determination of the cfDNA concentration beneficiates from determining two values determined by targeting two different WT sequences in two different genes (*BRAF* and *KRAS*). Samples with concentration values below a threshold revealing sample mishandling are excluded. The coefficient of variation of the cfDNA concentration from a blood sample is determined as 24 %. In each single run, negative and positive controls are quantified for each mutations and one standard curve is prepared. Q-PCR amplification is controlled by melt curve differentiation. RT-PCR assay is carried out in duplicate.

Detection of point mutations by Intplex® is very high since 1 copy (or 1 single molecule) out of 20,000 copies with WT respective sequence or out of one billion total cfDNA fragments can be detected in plasma sample (down to 2 copies/ml of plasma). Note that demonstration of the detection of one single molecule was made by analyzing concentration under compound Poisson distribution (probability distribution of the sum of a number of independent identically distributed random variables). Allele frequency range found in colorectal cancer patient plasma sample is 0.0013–0.2 %.

Short data turnaround time is another the assets of the Intplex technology as it is of 1–2 days (median) as observed in the Kplex1 study [23] and as well in the Kplex2 medico-economical study where samples were analyzed in the same conditions as in Molecular Biology Laboratory of various clinical centers where tumor tissue analysis data turnaround time is 16 days for RAS and BRAF testing (3 days to 3 months vs. 1 day to 10 days).

4.2 Melt Curve Analysis

In addition to DNA quantification and detection, confirmation of amplified sequence specificity by melt curve analysis appears as a necessary step when detecting in a clinical sample low-frequent mutations. Nevertheless, this step increases PCR analysis length. For instance, an approximately additional 1 h of machine time is needed if the fluorescence level is monitored from 50 to 90 °C with an increment of 0.2 °C. Surprisingly, the high-resolution melt additional step is a poorly explored strategy for the discrimination of mutant tumor cfDNA and healthy cfDNA. This approach corresponds to the high-resolution melting (HRM) curve analysis conventionally and widely used in many clinical center platforms for screening for unknown, low-abundance mutations and variants given its high sensitivity, rapidity, and cost-effectiveness.

The reported sensitivity of HRM is largely determined by fragment length, sequence composition, mutation identity, PCR quality, and equipment. Although recent publications report the ability to detect <1 % mutant in wild-type DNA, most applications of HRM-based assays exhibit a detection capability of approximately 5–10 % mutant among wild-type alleles. Although HRM mutation scanning is highly sensitive and efficient, HRM lacks the ability to identify the specific nucleotide change; this is a particularly important issue when mutations or variants are not known a priori and are likely to occur at any position within the amplicon sequence.

Melt curve analysis is not automatically available in recent dPCR technologies and in particular in ddPCR. In this case, the specificity of the primer amplification would need to be investigated before the experiment or based on the droplet repartition. With ddPCR, specific strategies should be employed in particular in case of multiplexing [21, 22].

4.3 Comparison of Intplex® with Methods Based on an Increased Sensitivity for the Detection of Genetic Alterations with cfDNA

Standard PCR-based assays have a relatively limited sensitivity for detecting infrequent genetic alterations and cannot detect mutations that represent less than 5–10 % of the total pool of alleles. Identification of somatic genetic and epigenetic aberrations now has been facilitated by the advent of highly sensitive technique. High-sensitivity polymerase is shown to induce less error during the amplification process. Tailored PCR focusing on specific structural genomic variants such as translocations, insertions, or deletions which are known to be present in the primary tumor allows for high sensitivity (0.001 %, which represents the detection of one mutant allele among 100,000 wild-type alleles). Point mutations consist one of the main somatic mutations especially in cancer patients and represent a technical challenge. Other qPCR-derived methods afford appropriate sensitivity for the detection of the point mutations. Nested PCR, ARMS, ASB, Bi-PAP, and allele-specific ligation PCR were developed [24, 25]. These methods are already used on clinical cohorts and validated for the detection of genetic alterations in tumor tissue samples.

In addition, new PCR amplification-based approaches include dPCR, beaming and Intplex. These PCR-based methods with a high sensitivity are limited by the fact that the exact genomic aberrations to be investigated must be known a priori. Beaming application appears limited since Beaming analysis must be outsourced due to the sophisticated technology and need of trained personnel. With its ability to perform absolute quantification of cfDNA, the dPCR technology should present a conceptual advantage compared to classical qPCR, especially for the determination of copy number variations. Whenever cfDNA is analyzed, copy numbers derived from PCR assay with fragmented DNA as a template cannot easily be interpreted as haploid whole-genome equivalents because the qPCR assay counts only the copies of the available full-length target. This implies that the detected copy number is, in fact, dependent on either the qPCR or the dPCR assay performed. Therefore, target copy numbers obtained for the same sample using different assays should only be compared after performing appropriate tests that check for quantification bias via a proper normalization. Digital droplet PCR translated to superior diagnostic performance such as in limiting interference by biological or chemical compounds on PCR and internal reproducibility but does not appear more sensitive as conventional Q-PCR when carried out under the same experimental conditions and appears as threefold more costly than Intplex®.

4.4 Validation of Intplex for the Determination of the Mutational Status of Known Hot-Spot Mutations

As regards to the determination of mutations on cfDNA Intplex® is limited to the mutational status of known hot-spot mutation; it is a "targeted approach." For instance, cfDNA analysis made possible the detection of the emergence of *RAS* and *BRAF* mutations following anti-EGFR therapy in mCRC patients: we validated the detection of *KRAS/BRAF* point mutation in 106 clinical samples from mCRC patients with 98 % of specificity with tumor tissue analysis in a blinded multicenter clinical study. Thus, we demonstrated the first clinical validation of the analysis of ctDNA in oncology [23].

4.5 Evaluation of Clinical Potential and Development of Intplex

Short data turnaround time is one of the assets of the Intplex technology as it is of 2 days (median) as observed in the Kplex1 study [23] and as well in the Kplex2 medico-economical study where samples were analyzed in the same conditions as in Molecular Biology Laboratory of various clinical centers where tumor tissue analysis data turnaround time is 15 days for RAS and BRAF testing (3 days to 3 months vs. 1 day to 10 days).

Intplex® is adaptable for types of genetic alterations (point mutation, deletion, exon skipping, copy number variation) for all types of cancer. Implementation in all clinical laboratories, i.e., as a kit, and the low data turnaround time and cost, makes this method well adapted for the longitudinal monitoring of cancer patients.

We will develop a new diagnostic technology to enable the delivery of personalized medicine in oncology by focusing on unmet clinical needs especially towards the longitudinal patient monitoring [24–26]:

1. Detecting minimal residual disease.
2. Monitoring patient treatment.
3. Early detection of drug resistance by testing mutation conferring drug resistance.
4. Detection of the minimal residual disease.
5. Surveillance of recurrence following treatment with curative intent.

Prospective blinded multicenter clinical investigations on those objectives are ongoing or planned for 2015. While the use of cfDNA analysis for monitoring residual disease has been evocated no demonstration has been made when using mutation load or the fragmentation index as a biomarker. In addition, this study may support claiming a way of combining total cfDNA and mutant ctDNA concentrations, mutation load, and the fragmentation index for increased synergistic power. An algorithm would be elaborated.

4.6 Perspectives of the Intplex® Technology

Our blinded prospective multicenter study clearly showed for the first time that ctDNA analysis could replace advantageously tumor section analysis. Procedures and pre-analytical conditions for analyzing circulating DNA were extensively studied and we propose an experimental guideline on those crucial points [19]; SOP is now urgently needed.

Intplex® is a patented technology [22] and can be adapted to all point mutations of interest in other genes in CRC or other malignancies. In addition to generating qualitative results on the mutation status for theragnostic purposes, cfDNA analysis by a method such as Intplex provides the possibility of dynamically determining mutation load, and it might further provide crucial information for monitoring treatment and for cancer patient follow-up. Furthermore, its use might bring light to the modification of tumor clonality upon its progression, and for the modification of mutant allele-specific imbalance (MASI) along with the copy number gain, which has been shown to be synergistically associated, noticeably in tumors harboring *KRAS* mutations.

The genomic appreciation of cancer leads to the reevaluation of the prevailing paradigm in clinical oncology in which the stages of clinical progression and anatomical observations govern the diagnostic and therapeutic principles [26]. As observed here, analyzing qualitatively and quantitatively genetic alterations with an efficient, simple, and cost-effective test from blood samples could expand the scope of the management of personalized cancer care.

Acknowledgments

A.R. Thierry is supported by the INSERM, France.

References

1. Mandel P, Metais P (1948) C R Seances Soc Biol Fil 142(3–4):241–243
2. Leon SA, Shapiro B, Sklaroff DM, Yaros MJ (1977) Free DNA in the serum of cancer patients and the effect of therapy. Cancer Res 37(3):646–650
3. Stroun M, Anker P, Lyautey J, Lederrey C, Maurice PA (1987) Isolation and characterization of DNA of cancer patients. Eur J Cancer Clin Oncol 23(6):707–712
4. Lo YMD, Chan KCA, Sun H, Chen EZ, Jiang PY, Lun FMF et al (2010) Maternal plasma DNA sequencing reveals the genome-wide genetic and mutational profile of the fetus. Sci Transl Med 2(61):61ra91
5. Fleischhacker M, Schmidt B (2007) Circulating nucleic acids (CNAs) and cancer – a survey. Biochim Biophys Acta 1775(1):181–232
6. van der Vaart M, Pretorius PJ (2008) Characterization of circulating DNA in healthy human plasma. Clin Chim Acta 395(1–2):186
7. Jung K, Fleischhacker M, Rabien A (2010) Cell-free DNA in the blood as a solid tumor biomarker--a critical appraisal of the literature. Clin Chim Acta 411(21–22):1611–1624
8. Schmidt B, Weickmann S, Witt C, Fleischhacker M (2008) Integrity of cell-free plasma DNA in patients with lung cancer and nonmalignant lung disease. Ann N Y Acad Sci 1137:207–213
9. Ellinger J, Wittkarnp V, Albers P, Perabo FGE, Mueller SC, von Ruecker A et al (2009) Cell-Free Circulating DNA: Diagnostic Value in Patients With Testicular Germ Cell Cancer. J Urol 181(1):363–371
10. Thierry AR, Mouliere F, Gongora C, Ollier J, Robert B, Ychou M et al (2010) Origin and

quantification of circulating DNA in mice with human colorectal cancer xenografts. Nucleic Acids Res 38(18):6159–6175

11. Mouliere F, Robert B, Peyrotte E, Del Rio M, Ychou M, Molina F et al (2011) High fragmentation characterizes tumour-derived circulating DNA. Plos One 6(9):e23418
12. Schwarzenbach H, Hoon DSB, Pantel K (2011) Cell-free nucleic acids as biomarkers in cancer patients. Nat Rev Cancer 11(6):426–437
13. Mouliere F, El Messaoudi S, Pang D, Dritschilo A, Thierry AR (2014) Multi-marker analysis of circulating cell-free DNA toward personalized medicine for colorectal cancer. Mol Oncol 8(5):927–941
14. Diehl F, Schmidt K, Choti MA, Romans K, Goodman S, Li M et al (2008) Circulating mutant DNA to assess tumor dynamics. Nat Med 14(9):985–990
15. Dawson SJ, Tsui DW, Murtaza M, Biggs H, Rueda OM, Chin SF et al (2013) Analysis of circulating tumor DNA to monitor metastatic breast cancer. N Engl J Med 368(13):1199–1209
16. Bettegowda C, Sausen M, Leary RJ, Kinde I, Wang Y, Agrawal N et al (2014) Detection of circulating tumor DNA in early- and late-stage human malignancies. Sci Transl Med 6(224):224ra24
17. Mouliere F, El Messaoudi S, Gongora C, Guedj AS, Robert B, Del Rio M et al (2013) Circulating cell-free DNA from colorectal cancer patients may reveal high KRAS or BRAF mutation load. Transl Oncol 6(3):319–328
18. Thierry AR, Molina F (2011) Analytical methods for cell free nucleic acids and applications, PCT WO 2012/028747, Sept 5
19. El Messaoudi S, Rolet F, Mouliere F, Thierry AR (2013) Circulating cell free DNA: preanalytical considerations. Clin Chim Acta 424:222–230
20. Bustin SA, Benes V, Garson JA et al (2009) The MIQE guidelines: minimum information for publication of quantitative real-time PCR experiments. Clin Chem 55(4):611–622
21. Taly V, Pekin D, El Abed A et al (2012) Detecting biomarkers with microdroplet technology. Trends Mol Med 18(7):405–416
22. Hindson CM, Chevillet JR, Briggs HA, Gallichotte EN, Ruf IK, Hindson BJ, Vessella RL, Tewari M (2013) Absolute quantification by droplet digital PCR versus analog real-time PCR. Nat Methods 10(10):1003–1005
23. Thierry AR, Mouliere F, El Messaoudi S, Mollevi C, Lopez-Crapez E, Rolet F et al (2014) Clinical validation of the detection of KRAS and BRAF mutations from circulating tumor DNA. Nat Med 20(4):430–435
24. Mouliere F, Thierry AR, Larroque C (2015) Detection of genetic alterations by nucleic acid analysis: use of PCR and Mass spectroscopy-based methods. In: Gahan P (ed) Circulating nucleic acids in early diagnosis, prognosis and treatment monitoring. Springer, Dordrecht
25. Misale S, Yaeger R, Hobor S, Scala E, Janakiraman M, Liska D et al (2012) Emergence of KRAS mutations and acquired resistance to anti-EGFR therapy in colorectal cancer. Nature 486(7404):532–536
26. Diaz LA, Williams RT, Wu J, Kinde I, Hecht JR, Berlin J et al (2012) The molecular evolution of acquired resistance to targeted EGFR blockade in colorectal cancers. Nature 486(7404):537–540

Chapter 2

COLD-PCR: Applications and Advantages

Zhuang Zuo and Kausar J. Jabbar

Abstract

Co-amplification at lower denaturation temperature-based polymerase chain reaction (COLD-PCR) is a single-step amplification method that results in the enhancement of both known and unknown minority alleles during PCR, irrespective of mutation type and position. This method is based on exploitation of the critical temperature, *Tc*, at which mutation-containing DNA is preferentially melted over wild type. COLD-PCR can be a good strategy for mutation detection in specimens with high nonneoplastic cell content, small specimens in which neoplastic cells are difficult to micro-dissect and therefore enrich, and whenever a mutation is suspected to be present but is undetectable using conventional PCR and sequencing methods. We describe in this chapter our COLD-PCR-based pyrosequencing method for *KRAS* mutation detection in various clinical samples using DNA extracted from either fresh or fixed paraffin-embedded tissue specimens.

Key words Polymerase chain reaction (PCR), Co-amplification at lower denaturation temperature PCR (COLD-PCR), *KRAS*, Mutation detection

1 Introduction

PCR works like a molecular photocopying machine. It makes millions of identical, amplified copies of a specific segment of DNA for downstream analysis. PCR is a key tool for DNA fingerprinting, forensic analysis, prenatal testing for genetic diseases, and finding mutations in cancer cells [1, 2].

PCR plays a vital role in detection of mutations in oncology specimens, most commonly when variant DNA sequences exist in the presence of a large majority of wild-type alleles as in case of heterogeneous tumors [3]. PCR does not contain an inherent selectivity towards variant (mutant) allele; thus both variant and non-variant alleles are amplified with approximately equal efficiency. The burden of identifying and sequencing a mutation in a PCR product falls on downstream assays. Despite being reliable for screening germline or somatic mutations, sequencing of unknown low-prevalence mutations using this otherwise powerful technology is still problematic.

Rajyalakshmi Luthra et al. (eds.), *Clinical Applications of PCR*, Methods in Molecular Biology, vol. 1392, DOI 10.1007/978-1-4939-3360-0_2,

With the rapid development of targeted therapies and personalized medicine, gaining knowledge of the genetic and molecular characteristics of a patient's tumor has become a crucial step in therapeutic and prognostic decision making in oncology. The specimens used for molecular testing range from large surgical resections to tiny fine-needle aspiration biopsies to cytology smears. Samples are submitted as either frozen or formalin-fixed, paraffin-embedded tissues, and the percentage of tumor cells varies substantially among different specimens [4].

One of the major challenges that clinical molecular diagnostics laboratories face is detection of mutations in samples with a low percentage of mutation-carrying tumor cells in the background of nonmalignant cells. The fact that the mutation detection limits of conventional sequencing techniques such as the gold standard Sanger sequencing is approximately 20 % and that of pyrosequencing is approximately 5–10 % suggests that we need introduction of tumor enrichment strategies for samples with a small tumor component [3, 5].

When tissue sections or smears on slides are used, tumor enrichment can be achieved at two levels: microscopically via manual microdissection or laser-capture microdissection and submicroscopically at the DNA level. Use of microdissection, one of the commonly used methods for tumor cell enrichment, requires substantial resources and expertise. Several sensitive methods have been developed to enrich specific mutations at the DNA level and bring detection limit down to 1 %. Examples of such methods include the PCR-based shifted termination assay and the non-PCR-based Invader assay [6, 7]; however, these assays require post-PCR manipulations and can be expensive. Methods such as denaturing HPLC and high-resolution melting curve (HRM) analysis enable detection of mutations, including unknown ones, at levels as low as 0.1 %. These low-level mutations, however, are not amenable to downstream conventional sequencing analysis [8, 9].

COLD-PCR is a new form of PCR that preferentially enriches "minority alleles" from mixtures of wild-type and mutation-containing sequences, irrespective of where a known or unknown mutation is present in the DNA sequence that is being interrogated [10–12]. COLD-PCR is 10–100 times more sensitive, depending on the specific DNA involved, than standard PCR for detecting genetic changes. This method utilizes a critical denaturation temperature to enrich unknown mutations at any position on the sequence, during amplification. The critical denaturation temperature (*Tc*) is lower than standard denaturation temperatures, and the method is usually applied for amplicons less than ~200 bp in length; hence in some cases more than one amplicon must be used to cover the full target region [11].

COLD-PCR can be used in any of the two formats, *full* COLD-PCR or *fast* COLD-PCR, depending on whether it is important to identify all possible mutations or achieve the highest mutation enrichment [10]. *Full* COLD-PCR enriches all possible mutations along the sequence, where an intermediate hybridization step temperature is used during PCR cycling to allow cross-hybridization of mutant and wild-type alleles (heteroduplexes). In *fast* COLD-PCR formation of heteroduplexes is not required and mutations are enriched anywhere along the sequence. The characteristic of both formats of COLD-PCR is to maintain adequate mutation enrichment for downstream sequencing, thus allowing the identification of the exact nucleotide change of low-prevalence mutations, though *fast* COLD-PCR has usually a greater mutation enrichment [12].

This method is very cost effective in terms of equipment and reagents and is relatively easy to implement. Studies have shown that COLD-PCR can substantially increase mutation detection sensitivity by 5- to 100-fold with different downstream detection methods, such as direct sequencing and restriction enzyme digestion [13].

In summary COLDPCR can be a good strategy for mutation detection in specimens with high nonneoplastic cell content, in small specimens in which neoplastic cells are difficult to microdissect and therefore enrich, and whenever a mutation is suspected to be present but is undetectable using conventional PCR and sequencing methods.

2 Materials

- Fresh or formalin-fixed and paraffin-embedded (FFPE) tissue samples.
- Genomic DNA extraction system, such as Autopure LS automated DNA extraction system (Qiagen, Valencia, CA) for fresh blood and bone marrow samples, or Qiagen QIAamp DNA FFPE Tissue Kit for FFPE samples.
- Oligonucleotide primers.
- AmpliTaq Gold polymerase (Life Technologies).
- PCR master mix:
  - Forward and reverse primers.
  - Deoxynucleotide triphosphate (dNTP) mix.
  - $MgCl_2$.
  - 20 mM Tris–HCl (pH 8.4).
  - 50 mM KCl.
- Thermal cycler, such as Applied Biosystems® 2720 Thermocycler (Life Technologies).
- Agarose gel equipment.
- Pyrosequencer (Qiagen) and reagents for downstream sequencing.

3 Methods

3.1 Overview of the General Principle

The principle of COLD-PCR is based on the observation that there is a critical denaturation temperature (*Tc*) for each DNA sequence that is lower than its melting temperature (*Tm*). PCR amplification efficiency for a DNA sequence drops abruptly if the denaturation temperature is set below its *Tc*. Even a single-nucleotide mismatch anywhere along the DNA sequence will generate a small but predictable change to the *Tm*. Depending on the sequence context and position of the mismatch, *Tm* can change 0.2–1.5 °C for a sequence of 200 bp. *Tc* is strongly dependent on DNA sequence, and it is determined empirically for each specific sequence. Figure 1 demonstrates the *Tm* profiles of the wild-type and mutant sequences at each of the six nucleotides of codons 12–13 of the *KRAS* gene. Studies have shown that PCR reaction yields no detectable amplification product when PCR denaturation temperature is set to below the *Tc*. The COD-PCR method takes advantage of this characteristic of PCR amplification to selectively enrich the minority alleles differing by one or more nucleotides at any position of a given sequence. In COLD-PCR, an intermediate annealing temperature is used during PCR cycling to allow cross-hybridization of mutant and wild-type alleles; heteroduplexes, which melt at lower temperatures than homo-duplexes, are then selectively denatured and amplified at *Tc*, while homo-duplexes remain double stranded and do not amplify efficiently. By fixing the denaturation temperature at *Tc*, mutations at any position along the sequence are enriched during COLD-PCR amplification.

3.2 COLD-PCR for KRAS Mutation Detection

Here we use *KRAS* mutation detection as an example to demonstrate the design of a COLD-PCR assay for clinical applications. A key consideration in designing a COLD-PCR assay that selectively amplifies the minority mutant alleles is to determine a new reduced denaturation temperature for the reaction. Ideally, this reduced denaturation temperature allows mainly the heteroduplexes to be denatured and amplified, and leaves the homo-duplexes double stranded and not amplified efficiently. We used the same primers as used in our conventional PCR assay, which produce an amplicon of 98 bp with 41.8 % of GC content and a *Tm* of 70.9 °C. Using the Poland algorithm, we plotted *Tm* profiles of the wild-type sequence along with those of mismatched sequences at each base pair within

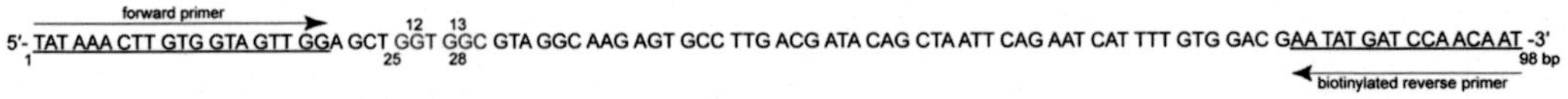

forward primer
12 13
5'- TAT AAA CTT GTG GTA GTT GGA GCT GGT GGC GTA GGC AAG AGT GCC TTG ACG ATA CAG CTA ATT CAG AAT CAT TTT GTG GAC GAA TAT GAT CCA ACA AT -3'
1 25 28 98 bp
biotinylated reverse primer

Fig. 1 Nucleotide sequence of the 98 bp PCR product of the *KRAS* gene and *Tm* profiles of the wild-type and mutant sequences at each of the 6 nucleotides of codons 12–13. *Arrows* indicate primer sequences (*underlined*). Sequence of codons 12–13 is indicated in *red*. Figure legends indicate *Tm* profiles for wild-type ("wild") and mutant ("m" and nucleotide number, e.g., "m25") sequences

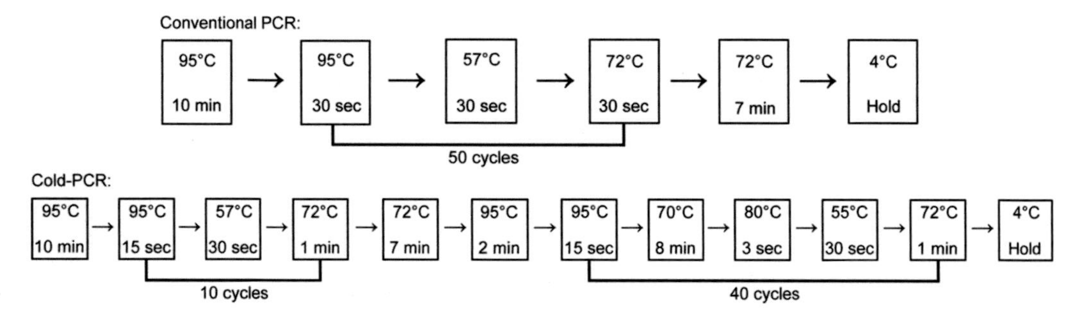

Fig. 2 Description of the PCR reaction conditions used in this study for conventional PCR and COLD-PCR protocols

KRAS codons 12 and 13 (Fig. 1). Based on this information, we set the reduced denaturation temperature of the COLD-PCR reaction at 80 °C. The reaction protocol started with 10 cycles of conventional PCR amplification for an initial buildup of all amplicons, followed by 40 COLD-PCR cycles to selectively enrich for mutant sequences. The initial conventional PCR cycling conditions are summarized as follows: 95 °C for 10 min; 10 cycles at 95 °C for 15 s, 57 °C for 30 s, and 72 °C, for 1 min; 72 °C for 7 min; and 95 °C for 2 min. Then 40 cycles of COLD-PCR were performed at 95 °C for 15 s, 70 °C for 8 min, 80 °C for 3 s, 55 °C for 30 s, and 72 °C for 1 min (Fig. 2). Full COLD-PCR reactions were performed on an ABI 2720 Thermocycler (Applied Biosystems). The COLD-PCR products were electrophoresed in agarose gels to confirm successful amplification of the 98 bp PCR product prior to pyrosequencing. As with our routine sequencing assay, a positive control, a negative control, and a reagent control were included in each run. All samples were run in duplicate.

3.3 Determining the Sensitivity of a COLD-PCR Assay

Sensitivity determination is a critical aspect of clinical assay development. A serial dilution study is usually performed for this purpose. For the *KRAS* COLD-PCR assay described above, we used a patient DNA sample containing a GGT-to-GCT mutation at codon 12 of the *KRAS* gene as the source of the mutant allele. This mutation-containing DNA sample was serially diluted with a wild-type DNA sample to obtain 1:2, 1:4, 1:8, 1:16, 1:32, 1:40, and 1:50 mutant to wild-type mixtures. All of the DNA mixtures were simultaneously subjected to conventional PCR and COLD-PCR followed by pyrosequencing. All samples were run in duplicate. The mutant-to-wild-type ratio was defined as the ratio of the peak height of a single mutant nucleotide over the peak height of a single wild-type nucleotide on pyrograms.

Figure 3 shows the pyrograms from the dilution study. With conventional PCR amplification, the mutant nucleotide peak on pyrogram became indistinguishable at 1:8 dilution, while it was clearly present at 1:32 dilution with COLD-PCR. Because the mutant sample we used for the dilution study was heterozygous at *KRAS* codon 12, a 1:32 dilution translates to about a 1.5 % of detection sensitivity, which is four times better than the 6 % sensitivity obtained with conventional PCR method.

As shown in Fig. 4, with increasing serial dilutions, the mutation-to-wild-type ratio decreased proportionally in a

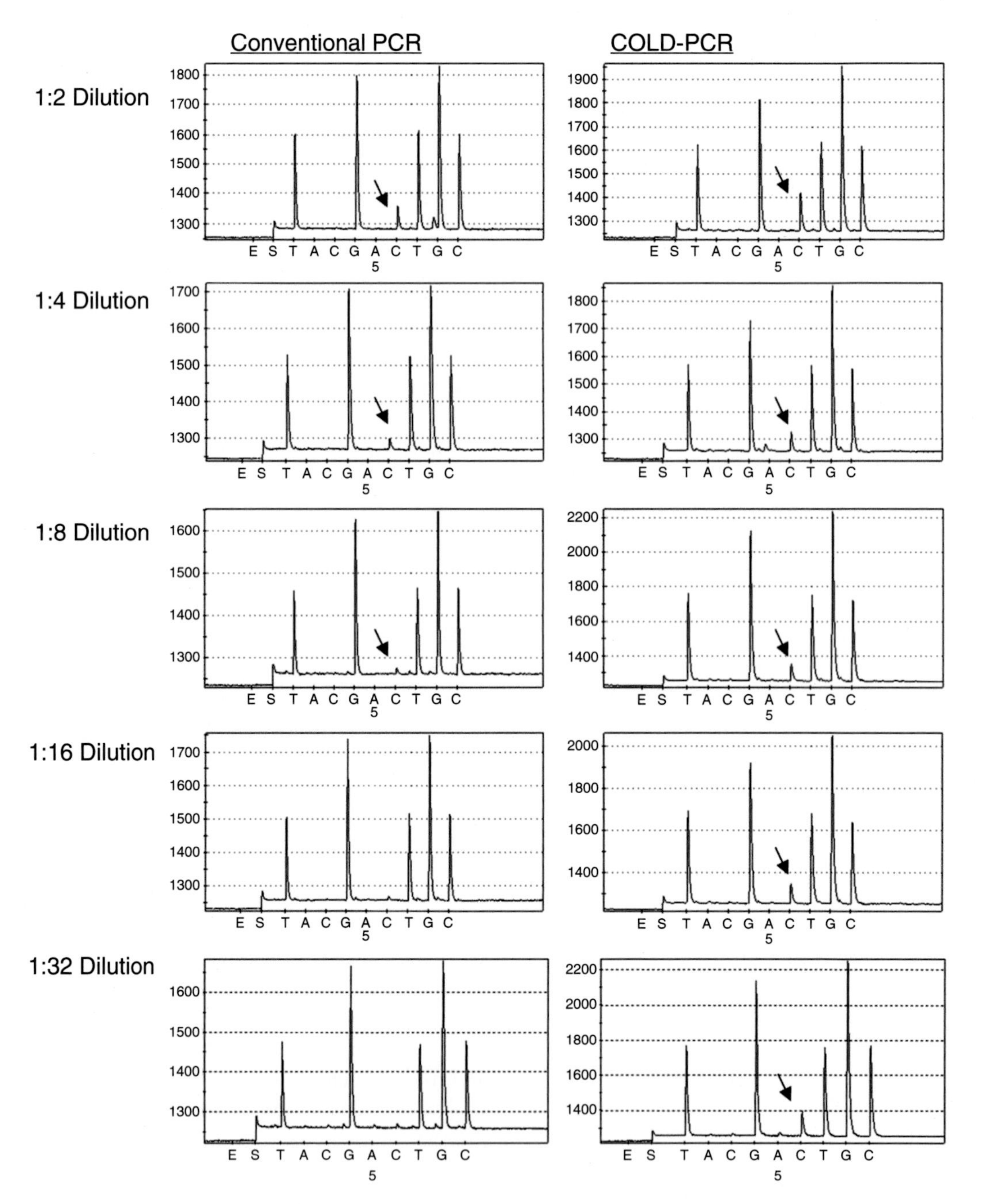

Fig. 3 Sensitivity determined by dilution study. Representative pyrograms of serial dilutions study with conventional PCR and COLD-PCR

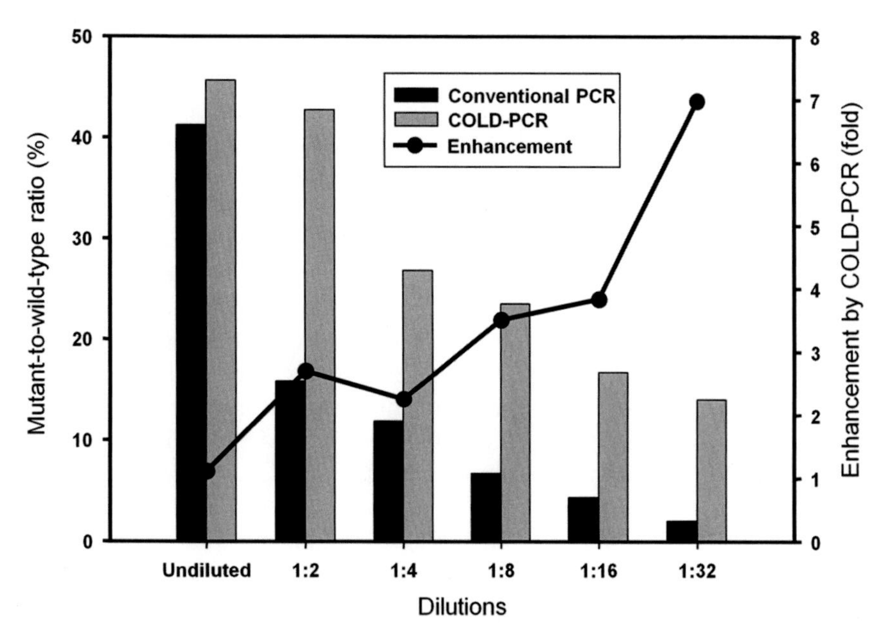

Fig. 4 Comparison of pyrosequencing results from undiluted to 1:32 dilution after conventional PCR and COLD-PCR (*bars*), and enhancement of COLD-PCR in serial dilutions (*line*)

first-order kinetics manner in both conventional PCR and COLD-PCR. The enhancement of mutation detection by COLD-PCR increased exponentially with increasing dilution. With the original specimen, which contains approximately 50 % mutant allele, COLD-PCR and conventional PCR had roughly the same efficiency in amplifying mutant alleles. At 1:32 dilution, COLD-PCR showed a sixfold greater efficiency in amplifying mutant allele compared with conventional PCR.

Within the detectable dilution range, all duplicate analyses of COLD-PCR products produced consistent pyrosequencing results. Different runs from the same sample also generated very similar results (correlation coefficient, 0.923).

4 Notes

1. COLD-PCR provides a general platform to improve the sensitivity of essentially all DNA variation detection technologies, such as Sanger sequencing, pyrosequencing, quantitative PCR, mutation scanning, genotyping, and methylation assays. As a general rule, a substantial enrichment for most COLD-PCR reactions can be obtained by using a *Tc* approximately 1 °C lower than the amplicon *Tm*; for certain sequences, however, fine-tuning of the *Tc* can be beneficial, and an optimal *Tc* can vary 0.5–1.5 °C lower than the *Tm*. The *Tm* can be experimentally determined on most real-time thermocyclers by performing a melting curve after PCR.

2. COLD-PCR can be applied in two formats, *full* COLD-PCR and *fast* COLD-PCR, depending on whether it is important to identify all possible mutations or to achieve the highest mutation enrichments. The example that we have demonstrated above is *full* COLD-PCR format. The *fast* COLD-PCR is a rapid PCR setup that skips the intermediate cross-hybridization step at 70 °C. For mutation enrichment to occur, the *full* COLD-PCR protocol requires the buildup of substantial PCR product to achieve efficient cross-hybridization, which restricts the enrichment to the late stages of PCR. In contrast, for *fast* COLD-PCR there is no requirement for PCR product buildup; hence the mutation enrichment starts at earlier PCR cycles than for *full* COLD-PCR. *Fast* COLD-PCR is rapid and results in higher enrichments than *full* COLD-PCR. However, in order to enrich for all possible mutations, including deletions/insertions, the *full* COLD-PCR program is necessary. By applying the *full* COLD-PCR program, a mismatch will always form between mutant and wild-type sequences and enrichment occurs irrespective of whether the specific nucleotide change tends to increase or decrease the *Tm*.
3. COLD-PCR selectivity for point mutations can increase further if subsequent PCR rounds are performed. As with deep sequencing approaches that use single-molecule sequencing, COLD-PCR enrichment of mutations is ultimately limited by polymerase-introduced errors. As newer polymerases with very high fidelity are continuously being improved, however, so are the ultimate enrichment abilities of approaches like COLD-PCR. Ultradeep sequencing following several rounds of COLDPCR could reveal aspects of cancer biology that are clinically very important (e.g., the origins of resistance to therapy).
4. Theoretical limitations of COLD-PCR: as amplicon size increases the difference between *Tm* and *Tc* decreases; amplification of polymerase-induced errors (preferentially use proofreader); requires optimization for each specific amplicon; well-to-well deviation from programmed temperature in many thermocyclers.

Acknowledgments

The authors thank Dr. Rajyalakshmi Luthra and Dr. L. Jeffrey Medeiros for supports of this work.

References

1. Lo YM, Corbetta N, Chamberlain PF et al (1997) Presence of fetal DNA in maternal plasma and serum. Lancet 350(9076):485–487
2. Kobayashi S, Boggon TJ, Dayaram T et al (2005) EGFR mutation and resistance of non-small-cell lung cancer to gefitinib. N Engl J Med 352(8):786–792
3. Li J, Makrigiorgos GM (2009) COLD-PCR: a new platform for highly improved mutation detection in cancer and genetic testing. Biochem Soc Trans 37:427–432
4. Luthra R, Zuo Z (2009) COLD-PCR finds hot application in mutation analysis. Clin Chem 55(12):2077–2078
5. Ogino S, Kawasaki T, Brahmandam M et al (2005) Sensitive sequencing method for KRAS mutation detection by pyrosequencing. J Mol Diagn 7:413–421
6. Yamamoto M, Kakihana K, Ohashi K et al (2009) Serial monitoring of T315I BCR-ABL mutation by Invader assay combined with RT-PCR. Int J Hematol 89(4):482–488
7. Li J, Wang L, Janne PA, Makrigiorgos GM (2009) Coamplification at lower denaturation temperature-PCR increases mutation-detection selectivity of TaqMan-based real-time PCR. Clin Chem 55(4):748–756
8. Nomoto K, Tsuta K, Takano T et al (2006) Detection of EGFR mutations in archived cytologic specimens of non-small cell lung cancer using high-resolution melting analysis. Am J Clin Pathol 126(4):608–615
9. Lin SY, Su YN, Hung CC et al (2008) Mutation spectrum of 122 hemophilia A families from Taiwanese population by LD-PCR, DHPLC, multiplex PCR and evaluating the clinical application of HRM. BMC Med Genet 9:53
10. Li J, Wang L, Mamon H et al (2008) Replacing PCR with COLD-PCR enriches variant DNA sequences and redefines the sensitivity of genetic testing. Nat Med 14(5):579–584
11. Milbury CA, Li J, Makrigiorgos GM (2009) COLD-PCR-enhanced high-resolution melting enables rapid and selective identification of low-level unknown mutations. Clin Chem 55(12):2130–2143
12. Mairinger FD, Vollbrecht C, Streubel A et al (2014) The "COLD-PCR approach" for early and cost-effective detection of tyrosine kinase inhibitor resistance mutations in EGFR-positive non-small cell lung cancer. Appl Immunohistochem Mol Morphol 22(2): 114–118
13. Zuo Z, Chen SS, Chandra PK et al (2009) Application of COLD-PCR for improved detection of KRAS mutations in clinical samples. Mod Pathol 22(8):1023–1031

Chapter 3

PCR-Based Detection of DNA Copy Number Variation

Meenakshi Mehrotra

Abstract

Copy number variations are important polymorphisms that can influence gene expression within and close to the rearranged region, and results in phenotypic variation. Techniques that detect abnormalities in DNA copy number are therefore useful for studying the associations between DNA aberrations and disease phenotype and for locating critical genes. PCR-based detection of copy number of target gene using TaqMan copy number assay offers a reliable method to measure copy number variation in human genome.

Key words Copy number variation (CNV), Array complete genomic hybridization (aCGH), PCR, TaqMan copy number assay, Copycaller software

1 Introduction

CNVs represent the most common form of structural genetic variation and their importance in genetic disease has been established [1–3]. CNVs can be caused by genomic rearrangements such as deletions, duplications, inversions, and translocations. There are different types of CNVs, from simple tandem duplications to more complex gains or losses of these sequences at multiple sites throughout the genome. A copy number variation occurs when a DNA segment of 1 kb to several megabases in length is present in variable copy numbers compared to a standard, reference genome [4]. The rapid progress in developing new technologies for the study of CNVs as array-comparative genomic hybridization (array-CGH), SNP arrays, and CNV-seq has led to an explosion in the amount of data now available [5–7]. aCGH has now become a standard tool in most clinical laboratories, and may soon even begin to replace conventional karyotyping as the primary method for investigating some disorders [8–11]. Arrays have an advantage of not dependent on the attainment of mitotically dividing cells within the tissue of investigation, as genomic DNA from tumor cells is used instead of metaphases. Despite this rapid growth, our understanding of the link between structural variants and human disease is somewhat

Rajyalakshmi Luthra et al. (eds.), *Clinical Applications of PCR*, Methods in Molecular Biology, vol. 1392,
DOI 10.1007/978-1-4939-3360-0_3,

limited. Technologies and statistical methods devoted to detection in CNVs from arrays inherent challenges in data quality associated with most hybridization techniques. Although the current versions of microarray-based techniques are sound, they are not the optimum platform for validation of copy number changes and high-throughput screening and it is desirable to have an alternative [5, 7, 12, 13]. A more quantitative technology offering a targeted approach, combined with high accuracy, specificity, ease of use, and sample throughput, is needed to validate copy number changes in large sample sizes. Quantitative real-time PCR based on TaqMan copy number assay offers a reliable alternative to measure copy number variation in human genome [14]. TaqMan copy number assays are run simultaneously with a TaqMan copy number reference assay in a duplex real-time PCR. The number of copies of the target sequence in each test sample is determined by relative quantitation (RQ) using the comparative Ct (ΔΔCt) method. Cycle threshold (Ct) is a measure of PCR cycles needed to get enough products to detect fluorescent signal (in this assay). This method measures the Ct difference (ΔCt) between target and reference sequences, and then compares the ΔCt values of test samples to a calibrator sample known to have two copies of the target sequence [14]. qPCR assays were performed using TaqMan Copy Number Assays (Applied Biosystems, Carlsbad, CA) for *c10orf 97*, *TP53* exon 5, exon 4 and exon 2, *c10orf 51*, and *RNF157* according to the manufacturer's instructions. The *RPP*40 gene, which is known to exist in two copies in a diploid genome, was used as the endogenous copy number reference in multiplex reactions. Healthy normal male genomic DNA was used as a diploid control.

2 Materials

2.1 Equipment

1. Microcentrifuge.
2. Vortex.
3. Pipets for PCR.
4. Filter tips.
5. Millipore water.
6. 1.5 ml microcentrifuge tubes.
7. Real-time PCR machine ABI 7900 HT.
8. ABI copy caller software.
9. MicroAmp optical reaction plate.
10. MicroAmp optical adhesive film (Applied Biosystems).

2.2 Reagents

1. Genomic DNA samples.
2. TaqMan genotyping master mix containing Ampli*Taq* Gold DNA polymerase, ultra-pure, containing dNTPs.

3. TaqMan copy number assay, *c10orf 97*, *TP53* exon 5, exon 4 and exon 2, *c10orf 51*, and *RNF157* 20× working stock (predesigned assays; Applied Biosystems) containing two unlabeled primers for amplifying the target sequence of interest.
4. TaqMan copy number reference assay, RNaseP/RPP40 20× (Applied Biosystems) (recommended standard reference assay for copy number quantitation experiments; this assay detects the ribonuclease P RNA component H1 (H1RNA) gene (RPPH1) on chromosome 14, cytoband 14q11.2).
5. Reference DNA with neutral copy number (human male DNA, Promega Corporation, Madison, WI).

3 Methods

3.1 Genomic DNA Extraction and Quantitation

The target template for TaqMan copy number assay is purified genomic DNA (gDNA). gDNA from bone marrow (BM) aspirates of CLL patients was isolated using the Autopure extractor (Qiagen/Gentra, Valenica, CA) according to the manufacturer's instructions and quantified using Nanodrop ND-1000 (Thermo Fisher Scientific, Wilmington, DE). 5 ng/μl diluted gDNA samples was used for further experimentation.

3.2 Setup of PCR Reaction

1. The PCR reactions were performed in triplicate using 20 ng of genomic DNA. Calculate the volumes of components required based on the reaction volume and the number of reactions as follows (include excess to provide for loss that occurs during reagent transfer):
   - 10 μl 2× TaqMan genotyping master mix.
   - 1.0 μl TaqMan copy number assay (20× working stock).
   - 1.0 μl TaqMan copy number reference assay (20× working stock).
   - 4.0 μl Millipore water.
   - 4.0 μl Sample genomic DNA.
   - 20.0 μl Total volume.
2. Completely thaw the TaqMan copy number assay mix for *c10orf 97*, *TP53* exon 5, exon 4 and exon 2, *c10orf 51*, and *RNF157* and TaqMan copy number reference assay mix. Gently vortex the assays to mix them, and then centrifuge the tubes briefly to bring contents to the bottom of the tube.
3. Mix the TaqMan genotyping master mix thoroughly by vortexing.
4. Combine the required volumes of reaction components in 1.5 ml microcentrifuge tubes.
5. Vortex and centrifuge the tubes briefly to mix the contents.

6. Pipet the reaction mix through multichannel pipettor into each of the wells of the reaction plate containing genomic DNA.
7. Seal the reaction plate with optical adhesive film (or optical caps), and then centrifuge.
8. Load the reaction plate into a real-time PCR instrument.
9. Run the plate using the parameters below:
   - Hold for 10 min at 95 °C.
   - 40 Cycles: 15 s at 95 °C.
   - 60 s at 60 °C.
   - In the "thermal cycler protocol" check "9600 Emulation" in the SDS 2.3 software on the 7900HT PCR system (Applied Biosystems as suggested by the manufacturer).
   - After the run is complete analyze the results in real-time PCR instrument using settings mentioned below:
   - Manual Ct: threshold: 0.2.
   - Autobaseline: On.
   - Review the analyzed data. Verify that the RPP40 reference assay and the copy number assays have a distinct amplification curve in all samples with a linear amplification phase.
   - Save the data in a .txt format and export it in Applied Biosystems CopyCaller Software v1.0 to analyze and determine the copy number for *c10orf 97*, *TP53* exon 5, exon 4 and exon 2, *c10orf 51*, and *RNF157* in each sample.

3.3 Result Analysis

Post-PCR copy number analysis of the target sequence in each test sample was performed by Applied Biosystems CopyCaller Software v1.0 (Applied Biosystems), which performs a comparative C_T ($\Delta\Delta C_T$) relative quantitation analysis of the real-time data using *RPP*40 and Promega normal male DNA as controls. CopyCaller® Software uses the target/detector names to distinguish the data from the different assays.

3.4 Analysis Settings in CopyCaller Software

Copy number analysis for each assay *c10orf 97*, *TP53* exon 5, exon 4 and exon 2, *c10orf 51*, and *RNF157* was performed using settings mentioned below (Fig. 1).

1. Mention the copy number of a normal male as 2 during analysis setting in calibrator selection panel.
2. Exclude samples with VIC C_T greater than 32.
3. Enter a ΔC_T threshold value as 4 to specify samples as zero copy samples.
4. Specify copy number bins (from 2 to 10) to estimate confidence.

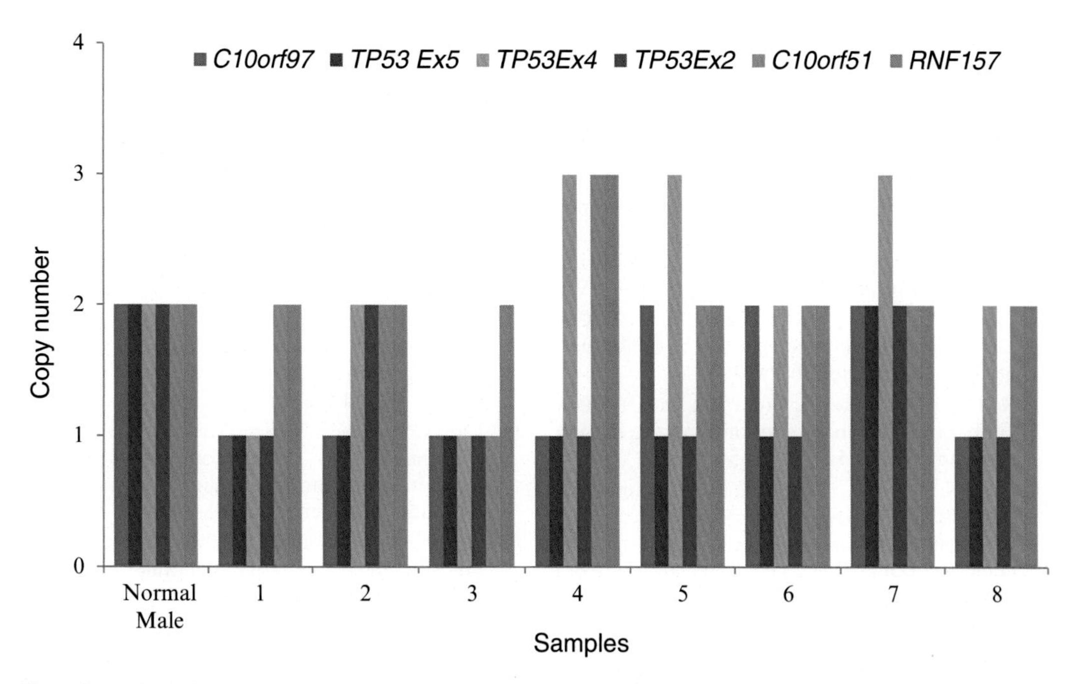

Fig. 1 Detection of copy number in eight cases of CLL by RT-PCR-based copy number analysis for *c10orf 97*, *TP53* exon 5, exon 4 and exon 2, *c10orf 51*, and *RNF157*. Single copy loss for 5 of 8 cases for *c10orf 97*, 7 of 8 cases for TP53 Ex5, 2 of 8 cases for TP53 Ex4, 6 of 8 cases for TP53 Ex2, 1 of 8 cases for *c10orf 51*, single copy gain for 3 of 8 cases for TP53 Ex4 and 1 of 8 tested cases for *c10orf 51* and *RNF157* whereas normal male did not show any copy number change for target genes tested

The software uses the statistical model $\Delta C_T = K - \log_{1+E}$ CN, where K is a constant, E is the PCR efficiency of the assay of interest, and CN is the copy number with the range $[1,\infty]$. Results of copy number assay can be reviewed in the form of copy number plot and table mentioning the copy number for each assay in individual sample. Copy number values are close to integers for each sample. Absolute Z score values and confidence values assess the reliability of each copy number call.

4 Notes

1. Concentration of genomic DNA should be uniform in each sample (Subheading 3.2, **step 1**).
2. A sample having "Undetermined" copy number if reference assay did not amplify sufficiently or replicate data is conflicting (Subheading 3.2, **step 8**; Subheading 3.4, **step 2**).
3. Samples may have a predicted copy number zero due to weak target amplification (Subheading 3.4, **step 4**).

References

1. Carson AR, Feuk L, Mohammed M et al (2006) Strategies for the detection of copy number and other structural variants in the human genome. Hum Genomics 2:403–414
2. Estivill X, Armengol L (2007) Copy number variants and common disorders: filling the gaps and exploring complexity in genome-wide association studies. PLoS Genet 3:1787–1799
3. Iafrate AJ, Feuk L, Rivera MN et al (2004) Detection of large-scale variation in the human genome. Nat Genet 36:949–951
4. Redon R, Ishikawa S, Fitch KR et al (2006) Global variation in copy number in the human genome. Nature 444:444–454
5. Iwao-Koizumi K, Maekawa K, Nakamura Y et al (2007) A novel technique for measuring variations in DNA copy-number: competitive genomic polymerase chain reaction. BMC Genomics 8:206
6. Vandeweyer G, Kooy RF (2013) Detection and interpretation of genomic structural variation in health and disease. Expert Rev Mol Diagn 13:61–82
7. Sharp AJ (2009) Emerging themes and new challenges in defining the role of structural variation in human disease. Hum Mutat 30:135–144
8. Shinawi M, Cheung SW (2008) The array CGH and its clinical applications. Drug Discov Today 13:760–770
9. Shaffer LG, Bejjani BA (2006) Medical applications of array CGH and the transformation of clinical cytogenetics. Cytogenet Genome Res 115:303–309
10. Rabin KR, Man TK, Yu A et al (2008) Clinical utility of array comparative genomic hybridization for detection of chromosomal abnormalities in pediatric acute lymphoblastic leukemia. Pediatr Blood Cancer 51:171–177
11. Oostlander AE, Meijer GA, Ylstra B (2004) Microarray-based comparative genomic hybridization and its applications in human genetics. Clin Genet 66:488–495
12. Bejjani BA, Shaffer LG (2006) Application of array-based comparative genomic hybridization to clinical diagnostics. J Mol Diagn 8:528–533
13. Sebat J, Lakshmi B, Troge J et al (2004) Large-scale copy number polymorphism in the human genome. Science 305:525–528
14. Mayo P, Hastshorne T, Li K et al (2010) CNV Analysis using Taq Man copy number assays. Curr Protoc Hum Genet 67:2.13.1–2.13.10

Chapter 4

Emulsion PCR: Techniques and Applications

Rashmi Kanagal-Shamanna

Abstract

Emulsion PCR (EmPCR) is a commonly employed method for template amplification in multiple NGS-based sequencing platforms. The basic principle of emPCR is dilution and compartmentalization of template molecules in water droplets in a water-in-oil emulsion. Ideally, the dilution is to a degree where each droplet contains a single template molecule and functions as a micro-PCR reactor. Here, we discuss the basic principles, advantages, and challenges of applications of emPCR in clinical testing. We describe the methods of preparation and enrichment of template-positive Ion PGM™ Template OT2 200 Ion Sphere™ Particles (ISPs) on the Ion Personal Genome Machine® (PGM™) System. For routine clinical testing, following library generation, we employ the automated Ion OneTouch™ System that includes the Ion OneTouch™ 2 and the Ion OneTouch™ ES instruments for template generation and enrichment of template-positive ISPs, respectively.

Key words Emulsion PCR, Ion PGM™, Ion Sphere™ Particles (ISPs)

1 Introduction

Targeted next-generation sequencing (NGS)-based assays facilitate high-throughput and cost-effective sequencing of genomic regions of interest in clinical diagnostics. As described elsewhere, the steps involved include library preparation, template generation and amplification, sequencing, and capture followed by data analysis [1]. This chapter specifically focuses on the use of emulsion PCR in "template generation and amplification" step for Ion PGM NGS-based assay.

Common to most commercially available platforms is DNA isolation and single-stranded library generation by random shearing of DNA into small fragments [2]. DNA fragments undergo adapter ligation and barcoding that are designed to facilitate further steps. Targeted selection/enrichment of genomic regions of interest (templates) from these randomly generated DNA fragments are performed by various methods including multiplex PCR, array-based, or solution-based hybridization [3]. These techniques have been elaborated in the chapter on PCR techniques in next-generation sequencing.

Rajyalakshmi Luthra et al. (eds.), *Clinical Applications of PCR*, Methods in Molecular Biology, vol. 1392,
DOI 10.1007/978-1-4939-3360-0_4, © Springer Science+Business Media New York 2016

In order to prepare the templates for sequencing, the templates have to be immobilized on a solid surface with sufficient amount of spatial separation to facilitate billions of simultaneous sequencing reactions [1]. This is the essential principle of massively parallel next-generation sequencing. The current methods of signal detection in most platforms routinely used in clinical laboratories (Ion PGM, 454, Illumina, and SOLiD) do not have the ability to detect single-base incorporation in an individual DNA molecule. Very few systems (Heliscope and PacBio SMRT systems) utilize unamplified single-molecule templates (amplification-free library) [2]. Thus, in most cases, the generated template molecules need to undergo a pre-clonal amplification step to enhance the signal-to-noise ratio to facilitate signal detection [4]. Emulsion PCR (EmPCR) is a commonly employed template amplification methodology in multiple NGS-based sequencing platforms including Ion PGM/Proton, Roche/454, Life/APG, SOLiD, and Polonator. Illumina/SOlexa use an alternative technology of solid-phase amplification (bridge amplification).

EmPCR is based on compartmentalization of DNA fragments in minute water droplets/vesicles in a water-in-oil emulsion to a degree of dilution where there is only a single or a few template molecules per droplet [5, 6]. Ideally, each vesicle/droplet contains one sphere, one single-stranded template molecule, one of the primers bound to the sphere, and all other reagents necessary for the PCR reaction; the second primer remains in the solution to screen out molecules bound to same adapters [4]. Thus, every vesicle functions as an isolated PCR micro-reactor leading to generation of numerous copies of the bound templates facilitating signal detection.

There are several advantages in using emPCR for the template amplification step in NGS-based platforms. PCR amplification of complex DNA libraries in NGS-based platforms is generally associated with inherent problems. These are primarily of two kinds: (1) preferential amplification of short fragments and (2) formation of chimeric DNA molecules by recombination of homologous regions [7]. These become especially important when using low-input template DNA, often the case in clinical diagnostics, as it requires higher number of PCR cycles. EmPCR is able to effectively mitigate these problems by separating the template molecules into numerous compartments, thereby preventing template competition and minimizing the chances of recombination [8, 9]. Specifically, since this step can heavily influence the selection of DNA molecules that are presented for sequencing, and consequently the data output [4], it is important to employ a robust technique optimized to produce minimal PCR amplification bias and thereby provide output DNA most representative of the original sample [1].

EmPCR is prone to general biases involved in any PCR techniques. In addition, due to the multiple steps involved in breaking of the emulsion and recovery and the double-Poisson distribution behavior, emPCR can result in suboptimal yield, thus requiring intensive optimization of PCR conditions [4]. Further, precise quantification of the input library to generate appropriate bead-to-fragment ratio is necessary to minimize generation of mixed signals from a single bead [10]. Extensive optimization is essential to minimize PCR amplification bias in order to provide the output DNA most representative of the original sample [1].

1.1 EmPCR for Template Generation and Amplification Step in Ion PGM and Proton (Life Technologies)

Ion PGM and Proton (Life Technologies) employ emPCR technology for template generation and amplification step. Ion PGM diagnostic assay is employed for assessment of mutational hotspot genomic regions in 50 known oncogenes and tumor-suppressor genes. The targeted panel yields 207 amplicons with an average length of 154 bp (111–187 bp). These amplicons undergo clonal amplification by emulsion PCR on the proprietary Ion Sphere™ Particles (ISPs), which is followed by sequencing. Here, we outline the methods of preparation and enrichment of template-positive Ion PGM™ Template OT2 200 ISPs for up to 200 base-read sequencing of a library on the Ion Personal Genome Machine® (PGM™) System. In our laboratory, following multiplex PCR, the stock DNA (library) is organized into sample pools that are then diluted to 7 pM concentration template generation and amplification step. For routine clinical testing, we employ the automated Ion OneTouch™ System that includes the Ion OneTouch™ 2 and the Ion OneTouch™ ES instruments for template generation and enrichment of template-positive ISPs, respectively. The steps are described under the following subheadings:

1. Preparation of template-positive ISPs containing clonally amplified DNA, using the Ion PGM™ Template OT2 200 Kit with the Ion OneTouch™ 2 instrument.
2. Enrichment of the template-positive ISPs with the Ion OneTouch™ ES instrument.

2 Materials

2.1 Preparation of Template-Positive ISPs Containing Clonally Amplified DNA, Using the Ion PGM™ Template OT2 200 Kit with the Ion OneTouch™ 2 Instrument

1. Pooled library (7 pM concentration).
2. Ion PGM Template OT2 Supplies 200 Kit containing:

   Ion OneTouch™ Reagent Tubes.

   Ion OneTouch™ Recovery Router.

   Ion OneTouch™ Recovery Tubes.

   Ion OneTouch™ 2 Amplification Plate.

   Ion OneTouch™ Sipper Tubes.

   Ion OneTouch™ ES (Enrichment) supplies.

   Ion OneTouch™ Lid.

   Ion OneTouch™ 2 Cleaning Adapters.

3. Ion PGM Template OT2 Solutions 200 Kit containing:

   Ion OneTouch™ Oil.

   Ion PGM™ Template OT2 200 PCR Reagent B.

   Ion OneTouch™ Reaction Oil.

   Nuclease-free water.

   Ion PGM™ OT2 Recovery Solution.

   Ion OneTouch™ Wash Solution.

   MyOne™ Beads Wash Solution.

   Neutralization solution.

   Tween® Solution.
4. Ion PGM Template OT2 Reagents 200 Kit containing:

   Ion PGM™ Template OT2 200 Reagent Mix (purple top).

   Ion PGM™ Template OT2 200 Enzyme Mix (brown top).

   Ion PGM™ Template OT2 200 ISPs (black top).
5. Dynabeads MyOne Streptavidin C1 Beads.
6. Eppendorf CombiTips Plus, 10 mL.
7. Eppendorf DNA LoBind Tubes, 1.5 mL.
8. Eppendorf microcentrifuge.
9. Pipettes (P10, P20, P200, P1000) and appropriate low-retention tips.
10. Vortexer with a rubber platform.

2.2 Enrichment of the Template-Positive ISPs with the Ion OneTouch™ ES

1. Ion PGM™ Template OT2 Solutions 200 Kit (Part no. 4481105) containing:

   Ion OneTouch™ Wash Solution (room temperature).

   MyOne™ Beads Wash Solution (room temperature, green top vial).

   TweenR Solution (room temperature).

   Neutralization solution (room temperature, red top vial).

   Nuclease-free water (room temperature).
2. Ion PGM™ Template OT2 Supplies 200 Kit (Part no. 4480981):

   8-Well strip.

   Eppendorf LoRetention Dualfilter Tips (P300).
3. Ion PGM™ Enrichment Beads (Cat no. 4478525).

   DynabeadsR MyOne™ Streptavidin C1 Beads.
4. 1.5 mL LoBind Tubes.
5. 0.2 mL PCR tubes.
6. 1 M NaOH.

7. Pipettes.
8. Vortexer.
9. DynaMag™-2 magnet.
10. Microcentrifuge.

3 Methods

3.1 Preparation of Template-Positive Ion PGM Template OT2 200 ISPs

1. Preparation of sample pools and 7 pM dilution:
   (a) Following target enrichment by multiplex qPCR, organize the generated samples (DNA stock) into pools based on the number of samples (*see* **Note 1**).
   (b) Dilute all samples to 200 pM concentration using appropriate amount of low TE (from Ion AmpliSeqTM Library Kit 2.0).
   (c) Use 2.0 μL of each sample to make a 20 pM pool dilution in low TE. Dilute the concentration further to 7 pM concentration to target 10–30 % of positive Ion PGM™ Template OT2 200 ISPs (*see* **Note 2**).
   (d) If required, the samples at various stages can be stored under the following conditions (Table 1).
2. Set up the Ion OneTouch™ 2 Instrument, and prepare the amplification solution using Ion PGM™ Template OT2 200 Reagent Kit as follows:
   (a) Completely thaw the reagent mix (purple top) at room temperature; vortex for 30 s and centrifuge for 2 s to ensure that there is no residual precipitate.
   (b) Vortex for 1 min and centrifuge for 2 s the PCR Reagent B for 1 min to ensure that the solution is clear and keep at room temperature.
   (c) Centrifuge (without vortexing) the enzyme (brown top) for 2 s, and keep on a cold block.

Table 1
Recommended storage conditions for samples at different stages of library dilutions

Sample type	Temperature (°C)	Storage duration
Library (DNA stock)	20	Forever
200 pM library dilution	20	6 Months
Pooled dilution	4	1 Month

Table 2
Components of Ion PGM™ Template OT2 200 Reagent Kit to prepare amplification solution

Order	Reagent	Volume (μL)
1	Nuclease-free water	25
2	Ion PGM™ Template OT2 200 Reagent Mix (Purple cap)	500
3	Ion PGM™ Template OT2 200 PCR Reagent B (Blue cap)	300
4	Ion PGM™ Template OT2 200 Enzyme Mix (Brown cap)	50
5	Diluted library (not stock library) 7 pM	25

(d) Add the following components in the designated order to a 1.5 mL LoBind Tube at 15–30 °C (Table 2). Pipette the amplification solution after adding each component to ensure adequate mixing.

(e) Vortex the amplification solution at maximum speed for 5 s, and then centrifuge the solution for 2 s.

3. Vortex the Ion PGM™ Template OT2 200 ISPs at maximum speed for 1 min to resuspend the particles, and then pipette for adequate mixing; then, add 100 μL of ISPs to the amplification solution. Vortex this at maximum speed for 1 min.

4. Fill the Ion PGM OneTouch™ Plus Reaction Filter Assembly:

   (a) Place the Ion PGM OneTouch™ Plus Reaction Filter Assembly into a tube rack with the three ports facing up. Identify the sample port, which is the one connected to the short tubing in the reaction tube.

   (b) Through the sample port, add 1000 μL of amplification solution (after vortexing at maximum speed for 5 s, centrifuging for 2 s, and mixing by pipetting up and down) through the sample port (aqueous phase) (*see* **Note 3**) followed by 1500 μL of Ion OneTouch™ Reaction Oil (oil phase). The aqueous and the oil phase should be distinguishable in the reaction tube.

   (c) Install the filled Ion PGM™ OneTouch Plus Reaction Filter Assembly on the Ion OneTouch™ 2 Instrument:

   - Rotate the Ion PGM™ OneTouch Plus Reaction Filter Assembly in one sweep, until the sample port (facing up) on the left is inverted and faces down along with the other ports. This is critical to minimize the contact of the amplification solution to the short tubing in the reaction tube.
   - Insert the three ports into the three holes on the top stage of the instrument.

5. Remove the samples ≤16 h after starting the run.
6. Recover the template-positive Ion PGM™ Template OT2 200 ISPs:
   (a) Remove both the Ion OneTouch™ Recovery Tubes from the instrument; remove the supernatant without disturbing the ISP pellet.
   (b) Resuspend the pellet in the remaining Ion PGM™ OT2 Recovery Solution and transfer the suspensions from both the tubes into a 1.5 mL LoBind Tube for a total of 100 μL.
   (c) Again, rinse each recovery tube with 125 μL of Ion OneTouch™ Wash Solution and transfer all 250 μL to the 100 μL ISP suspension in the LoBind Tube (*see* **Note 4**).
   (d) Centrifuge the ISPs at 15,500 × *g* for 2.5 min. Remove all of supernatant except 100 μL and resuspend the ISPs in 100 μL of Ion PGM™ OT2 Wash Solution (*see* **Note 5**).
   (e) 2 μL of the solution is used for Qubit quality control testing; rest of the sample is used for the subsequent enrichment step.

3.2 Enrichment of the ISPs Using the Ion OneTouch™ ES Instrument

1. Prepare fresh melt-off solution by combining in the order indicated in Table 3 (*see* **Note 6**).
2. Prepare the Dynabeads® MyOne™ Streptavidin C1 Beads:
   (a) Vortex the Dynabeads® MyOne™ Streptavidin C1 Beads for 30 s and transfer 13 μL to a new 1.5 mL LoBind Tube.
   (b) Place the tube on a magnet such as a DynaMag™-2 for 2 min; discard the supernatant without disturbing the bead pellet.
   (c) Add 130 μL of MyOne™ Beads Wash Solution.
   (d) Remove the tube from the magnet, vortex the tube for 30 s, and centrifuge the tube for 2 s.
3. Fill the wells in the 8-well strip designed for the ES instrument as indicated in Table 4. Well 1 is closest to the square-shaped tab.

Table 3
Preparation of melt-off solution for the enrichment of the Ion Sphere™ Particles

Order	Component	Volume (mL)
1	Tween solution	280
2	1M NaOH	40
	Total	320

Table 4
Allotment of the reagents in the wells of the 8-well strip designed for the ES instrument

Well number	Reagents to dispense in the well
Well 1	Entire ISP sample (100 μL)
Well 2	130 μL of DynabeadsR MyOne™ Streptavidin C1 Beads resuspended in MyOne™ Beads Wash Solution (B)
Well 3	300 μL of Ion OneTouch™ Wash Solution (W)
Well 4	300 μL of Ion OneTouch™ Wash Solution (W)
Well 5	300 μL of Ion OneTouch™ Wash Solution (W)
Well 6	Empty
Well 7	300 μL of freshly prepared melt-off solution (M)
Well 8	Empty

4. Insert the filled strip into the machine slot with the square-shaped tab on the left, and the strip is pushed all the way to the right end of the slot.
5. Load a new tip in the Tip Arm of the instrument.
6. Add 10 μL of the neutralization solution to a new 0.2 mL PCR tube. Load the opened tube into the hole in the base of the Tip Loader.
7. Pipette the contents of Well 2 up and down to resuspend the beads without introducing bubbles.
8. Turn the instrument on and set the timer on for 35 min.
9. After 35 min when the run is completed, return immediately to close and remove the 0.2 mL PCR tube (containing the enriched ISPs). Mix the contents of this tube by gently inverting it about five times. The tube may also be vortexed and spun down alternatively.
10. There are three things that need to be checked at the end of the run:
    (a) The 0.2 mL PCR tube must have >200 μL of solution containing the enriched ISPs. If not, *see* **Note 7**.
    (b) It is important to make sure that the Well 1 contains only residual MyOne beads (≤2 μL) at the end of the run. Well 2 also contains a visible amount of MyOne beads. Wells 2–5 and 7 contain waste or residual unused reagents.
    (c) It is important to make sure that the ISP pellet is not tinted brown. If so, *see* **Note 8**.

Table 5
Acceptance criteria for the unenriched Ion Sphere™ Particle (ISP) samples based on the percent templated ISPs determined by Qubit 2.0 Fluorometer

Percent templated ISPs (%)	Interpretation
<10	Sample contains an insufficient number of templated ISPs to achieve optimal loading density on the Ion chip
10–30	Optimal amount of library
>30	Sample will yield multi-templated ISPs (mixed reads)
>10	Acceptable for sequencing as long as the raw values are >100 RFUs

11. Enriched ISPs can be sequenced using Ion PGM™ Sequencing 200 Kit. These can be stored at 2–8 °C for up to 3 days.
12. For unenriched ISP samples, determine the percent templated ISPs using Qubit 2.0 Fluorometer (beyond the scope of this chapter). The optimal amount of library corresponds to the library dilution point that gives percent templated ISPs between 10 and 30 %. If the results are outside the desired percent templated ISP range, the library input can be altered appropriately. The acceptance criteria are listed in Table 5.

4 Notes

1. In our lab, we use 8 patient samples and 1 control library to make up one pool when using the 318Cv2 chip. We use 3 patient samples and 1 control library to make up one pool when using a 316 chip. If the number of samples with adequate library concentrations is not a multiple of 8 or 3, we divide the samples equally between the pools. It is important to note that no pool should have the same sample barcode.
2. Here is an illustration to make a pool for 318Cv2 chip (9 samples): To a labeled 1.5 mL Lo-Bind Tube, add 162 μL of Low TE + 2.0 μL from each of the 200 pM dilutions made to get a total volume of 190 μL. The new concentration of the pool will be 20 pM overall. To make a 7 pM pool, take 35 μL of 20 pM pool and add 65 μL of Low TE. The 7 pM pool will be used to set up emulsion PCR in the subsequent step.
3. Insert the pipette tip into the sample port perpendicularly and firmly to form a tight seal. Slowly load the 1000 μL of the amplification solution through the sample port. Avoid aspirating solution from the Ion PGM™ OneTouch Plus Reaction Filter Assembly. After discarding the tip, if necessary, gently dab a Kimwipes® disposable wiper around the ports to remove any liquid.

4. Recovered Ion PGM™ Template OT2 200 ISPs can be stored at 2–8 °C for up to 3 days. These can be processed similar to freshly recovered ISPs by centrifuging at 15,500 × *g* for 2.5 min, removing all but 100 μL of supernatant, and resuspending them in wash solution.
5. If the volume of templated ISPs <100 μL, add sufficient wash buffer to bring the volume to 100 μL.
6. Melt-off solution must be prepared daily.
7. After a successful run, the 0.2 mL PCR tube must have ~230 μL of the Ion OneTouch™ Wash and Neutralization solutions containing enriched ISPs. If the 0.2 mL tube has <200 μL of solution, check the Well 8 for ISPs. If there are ISPs in Well 8, transfer the ISPs from Well 8 to the PCR tube, then re-calibrate the Ion OneTouch ES, and perform a residual volume test. Prolonged exposure to the Melt-Off solution can cause ISP and DNA damage.
8. A brown-tinted pellet indicates that the MyOne Beads are mixed with the ISPs. In that case, resuspend the pellet in the 0.2 mL PCR tube and place against a magnet (DynaMag-2 magnet) for 4 min. Transfer the ISPs once again to a new tube without the pellet.

References

1. Metzker ML (2010) Sequencing technologies – the next generation. Nat Rev Genet 11:31–46
2. Gullapalli RR, Desai KV, Santana-Santos L et al (2012) Next generation sequencing in clinical medicine: challenges and lessons for pathology and biomedical informatics. J Pathol Inform 3:40
3. Mamanova L, Coffey AJ, Scott CE et al (2010) Target-enrichment strategies for next-generation sequencing. Nat Methods 7:111–118
4. Buermans HP, den Dunnen JT (2014) Next generation sequencing technology: advances and applications. Biochim Biophys Acta 1842(10):1932–1941
5. Nakano M, Komatsu J, Matsuura S et al (2003) Single-molecule PCR using water-in-oil emulsion. J Biotechnol 102:117–124
6. Dressman D, Yan H, Traverso G et al (2003) Transforming single DNA molecules into fluorescent magnetic particles for detection and enumeration of genetic variations. Proc Natl Acad Sci U S A 100:8817–8822
7. Meyerhans A, Vartanian JP, Wain-Hobson S (1990) DNA recombination during PCR. Nucleic Acids Res 18:1687–1691
8. Williams R, Peisajovich SG, Miller OJ et al (2006) Amplification of complex gene libraries by emulsion PCR. Nat Methods 3:545–550
9. Xuan J, Yu Y, Qing T et al (2013) Next-generation sequencing in the clinic: promises and challenges. Cancer Lett 340:284–295
10. Yu B (2014) Setting up next-generation sequencing in the medical laboratory. Methods Mol Biol 1168:195–206

Chapter 5

Digital PCR: Principles and Applications

Rashmi Kanagal-Shamanna

Abstract

Digital PCR is a robust PCR technique that enables precise and accurate absolute quantitation of target molecules by diluting and partitioning of the samples into numerous compartments. Automated partitioning can be attained by creating "water-in-oil" emulsion (emulsion-based digital PCR) or using a chip with microchannels (microfluidics-based digital PCR). We discuss the advantages and a wide variety of clinical applications of this technique. We describe the droplet digital RT-PCR protocol published by Jennings et al. for identification and absolute quantitation of BCR-ABL1 transcripts.

Key words Droplet digital PCR, Emulsion PCR, Microfluidics, BCR-ABL1 transcript

1 Introduction

Digital PCR is a robust PCR technique that enables accurate absolute quantification of target molecules at a high degree of sensitivity. It essentially combines the simplicity of traditional end-point PCR and the quantification features of the real-time quantitative PCR (qPCR) methodologies. Unlike qPCR, the quantitation is absolute and does not utilize calibration of standards, thus making the process faster, more precise, and reproducible [1].

The basic principle involves extreme limiting dilution and partitioning of the sample into millions of separate units that ideally contain either no particle or a single particle. Each unit contains all the reagents necessary for a PCR reaction, and basically functions as a micro-PCR reactor. If the unit contains the template of interest, PCR amplification yields a positive signal. If there is no template, there is no signal. The quantitation is binary, hence the term "digital." If the number of partitioning units is known, the initial amount of target molecules can be estimated by the knowledge of the total number of positive to negative signals [2]. At an extreme dilution, this simply represents the ratio of the positive to negative units as there is either no template or a single template within the unit. At a higher concentration, to account for the possibility of

Rajyalakshmi Luthra et al. (eds.), *Clinical Applications of PCR*, Methods in Molecular Biology, vol. 1392,
DOI 10.1007/978-1-4939-3360-0_5, © Springer Science+Business Media New York 2016

multiple target molecules in each unit, an adjustment factor using Poisson distribution analysis is incorporated into the analysis. In such a case, the number of target molecules can be determined using the equation $\lambda = -\ln(1 - p)$, where λ is the average number of target molecules per unit and p is the ratio of the number of units with positive signal to the total number of partition units. The absolute target concentration can be estimated using λ, the reaction volume in each of the units, and the total number of partition units. It is apparent that the larger number of partitions, the higher the precision and the dynamic range of the assay [3].

Traditionally, digital PCR was an elaborate procedure that required conducting serial dilutions of the sample followed by PCR amplifications. Currently, advanced techniques can be conveniently used for automated partitioning of the sample of interest into numerous (up to 100,000) known number of separate micro-PCR reactors designed to amplify the target of interest [4]. Most instruments commonly employ one of the two modalities of partitioning. The first is based on emulsion chemistry which involves creating emulsions of aqueous droplets in an oil medium, where each droplet functions as a PCR unit (known as droplet digital PCR, ddPCR) [5, 6]. The second involves microfluidics where a physical microchannel (either surface or capillary based) system is designed on plates or arrays to create partitions [7, 8]. Emulsion-based digital PCR systems allow analysis of a greater range of sample concentrations for any given precision than chip-based systems due to a much larger number of partitions [4].

There are several distinct advantages of digital PCR compared to other types of PCR techniques. Digital PCR can detect target molecules at a much lower frequency (in 100,000) than qPCR [3]. In fact, the sensitivity in this is limited only by the number of molecules present in the sample that can be amplified [5]. With regard to copy number analysis, it has been shown that changes less than 1.2-fold difference can also be detected by digital PCR [9]. As mentioned before, digital PCR provides an absolute quantitation of nucleic acids without the need for an external standard, thus making it faster, more precise, and reproducible than qPCR [2]. Since quantitation involves determination of the presence or absence of amplification signal in each of the multiple units, similar to an end-point PCR, it is less dependent on the factors affecting PCR amplification efficiency such as the presence of inhibitors, target/primer accessibility issues due to poor denaturing, or point mutation on the primer or probe annealing site of the target molecule [10]. Digital PCR technique facilitates easier design of multiplexing for assessment of multiple targets. Use of traditional multiplex qPCR can be a challenge for measurement of multiple targets when there is marked difference between various target concentrations leading to preferential amplification of high-abundance target molecules by depletion of reagents. Digital PCR overcomes this challenge since each individual micro-reaction is generally positive for one of the templates. This has been aptly termed "synthetic enrichment effect" [10].

Digital PCR has already been employed in a variety of applications in cancer diagnostics that include detection of rare mutations [3, 7, 8], copy number analysis, specifically for HER2 and EGFR [3, 9, 11–16], as well as absolute quantitation [17–20]. In addition, it has been widely used in microbiological identification and quantification of bacteria [21] and viruses [22–24]. Further, digital PCR has been shown to be more sensitive than qPCR in cell-free fetal DNA testing for paternally inherited single-gene disorders [25, 26]. For efficient adoption and utilization of this technology by the scientific community in general, minimum experimental parameters to be indicated in the protocols are available [4].

Jennings et al. have shown that the droplet digital RT-PCR offers a much simpler interface for identification and quantitation of BCR-ABL1 transcripts, with an improved lower limit of detection and absolute quantification without bias [18]. Using their technique, the sample is partitioned into 20,000 nanodroplets that are individually amplified by PCR. This is followed by identification of the number of positive and negative droplets for BCR-ABL1 fusion transcript, BCR transcript, and both by fluorescence detectors. Here, we describe the protocol used by the authors for identification of BCR-ABL1 transcripts using digital PCR and based on the recommendations from the manufacturer (Bio-Rad). Briefly, the steps involved include cDNA generation, droplet generation, PCR, droplet reading, and analysis.

2 Materials

1. Primers and probes.

   TaqMan MGB probes and primers (Life Technologies, Carlsbad, CA) generated for the analysis are as follows: BCR-ABL1 (forward) 5′-CATTCCGCTGACCATCAATA-3′; BCR-ABL1 (reverse) 5′-ACACCATTCCCCATTGTGAT-3′; BCR-ABL1 probe 5′-/56-FAM/CCCTTCAGCGGCCAGTAGC ATCTGA-3′; BCR (forward) 5′-CAGTGCGTGGAGGAGA TCGA-3′; BCR (reverse) 5′-CGATGCCCTCTGCGAAG TTG-3′; and BCR probe 5′-VIC-CAGCCTTCGACGTCAA-3′.
2. Input RNA.
3. Nuclease-free water.
4. One-Step RT-ddPCR kit from Bio-Rad Laboratories (Cat# 1863021; Hercules, CA).
5. Heat block *or* water bath.
6. 96-Well PCR plates.

7. Pipets.
8. Pipet tips.
9. Centrifuge.
10. QX100 droplet generator (Bio-Rad).
11. DG8 droplet generator cartridges (single-use; Bio-Rad).
12. DG8 droplet generator cartridge holder (Bio-Rad).
13. DG8 gaskets (single-use; Bio-Rad).
14. ddPCR droplet generation (DG) oil (Bio-Rad).
15. Bio-Rad QX100 droplet reader.
16. Droplet reader plate holders.
17. QuantaSoft software.
18. Eppendorf twin.tec semi-skirted 96-well plate.
19. Heat sealer.
20. Heat sealing plate foil.
21. Thermal cycler.

3 Methods (Adapted from Jennings et al. [18])

1. Incubate the RNA samples for 5 min at 75°.
2. Thaw the components of the One-Step RT-ddPCR kit from Bio-Rad Laboratories (Cat# 1863021; Hercules, CA) at room temperature. The contents of each of the reagents are adequately mixed by either pipetting or inverting followed by centrifugation.
3. Prepare the RT-PCR reaction mixtures for BCR and BCR-ABL1 transcripts as detailed in Tables 1 and 2.

Table 1
RT-ddPCR reaction mix for BCR transcript

Component	Volume per reaction	Final concentration
2× One-step RT-ddPCR supermix	10 μL	1×
25 mM manganese acetate solution	0.8 μL	1 mM
Forward BCR primer		900 nmol/L
Reverse BCR primer		900 nmol/L
BCR probe		250 nmol/L
RNase/DNase-free water		
RNA template	40 ng	
Total volume	20 μL	

Table 2
RT-ddPCR reaction mix for BCR-ABL1 transcript

Component	Volume per reaction	Final concentration
2× One-step RT-ddPCR supermix	10 μL	1×
25 mM manganese acetate solution	0.8 μL	1 mM
BCR-ABL1 forward primer		2250 nmol/L
BCR-ABL1 reverse primer		2250 nmol/L
BCR-ABL1 probe		625 nmol/L
RNase/DNase-free water		
RNA template	600 ng	
Total volume	20 μL	

4. Dispense equal amounts of the reaction mixtures to each of the reaction tubes and then add appropriate amounts of the template. Mix either by pipetting or vortexing followed by centrifugation and let it stand at room temperature.
5. Insert the DG8 cartridge into the holder. The notch in the cartridge should be located at the upper left side of the holder.
6. Pipet 20 μL of droplet digital PCR (ddPCR) reaction mixture from eight samples at a time (make sure it is at room temperature) using eight-channel pipet and load it into wells designated as "sample" (middle row) on the Bio-Rad DG8 disposable droplet generator cartridge. It is important to avoid air bubbles while pipetting that may interfere with the droplet generation (*see* **Note 1**).
7. Pipet 70 μL of droplet generation oil into each of the wells designated "oil" (bottom row) for each of the samples on the cartridge.
8. All eight samples designated "samples" must contain either sample of 1× buffer control. All eight samples designated oil should contain droplet generator oil.
9. Firmly attach a new rubber gasket over the cartridge holder (*see* **Note 2**).
10. Place the cartridge holder in the QX100 droplet generator, and close the lid.
11. Droplet generation is completed in about 2 min as indicated by all the three lights on the machine turning solid green.
12. Remove the holder with the cartridge still in place from the QX100. Removing the gasket, and inspect. The sample and oil wells should be almost empty. The droplets should be present in the "droplet wells" (top row) and should be slightly opaque.

13. Transfer 40 μL from each of the wells with generated droplets gently to an Eppendorf96-well twin.tec PCRplate (Eppendorf, Hamburg, Germany) using multichannel micropipette (*see* **Note 3**).
14. Immediately cover the contents of the wells to prevent evaporation.
15. After all the samples have been loaded, heat-seal the plate with a Thermo Scientific Easy Pierce pierceable foil seal from Thermo Fisher Scientific (Cat#AB-0757; Waltham, MA). Avoid centrifugation once the droplets are generated.
16. The cartridge can then be discarded from the holder.
17. PCR should be performed within half hour of sealing the plate. If not, the plate can be stored for up to 4 h at 4 °C. The plate is placed on a thermal cycler and amplified to the end-point at conditions indicated below:

    60 °C × 30 min (1 cycle), 95 °C × 5 min (1 cycle), 95 °C × 30 s (ramp rate 2.5 °C/s), and 59 °C × 60 s (ramp rate 2.5 °C/s) (40 cycles), 98 °C × 10 min (1 cycle), and a 12 °C hold.
18. Secure the PCR plate on the PCR plate holder.
19. Turn on the QX100 droplet reader and launch the QuantaSoft analysis software version 1.3.2.0. It is important to make sure that the droplet reader has sufficient droplet reader oil in the supply bottle and less than 700 mL in the waste bottle.
20. Load the 96-well PCR plate on the holder into the Bio-Rad's QX100 droplet reader and close the lid.
21. Set up a new experiment with the sample details on the QuantaSoft analysis software and start the run. Individual droplets are analyzed as positive and negative by a two-color detector to provide absolute quantification in digital form. Remove the PCR plate and the holder and discard the PCR plate.
22. Once the run is completed, data can be analyzed.

4 Notes

1. According to the Bio-Rad manual, air bubbles, even when not visible, prevent the samples from reaching the bottom of the wells and into the microchannels, thereby significantly reducing the amount of droplets. They recommend the use of 20 μL aerosol-barrier (filtered) Rainin pipet tips and not 200 μL pipette tips. Instead of pipetting the sample directly on the side of the well, they recommend gently sliding the pipet down the side wall at a 15° angle past a ridge near the bottom. Gently pipet half the sample at 15° angle at this location, and dispense the rest while sliding the pipette back up the wall.

2. The gasket must be securely attached to ensure sufficient pressure for droplet generation.
3. For transferring the droplets from the cartridge onto the PCR plate, Bio-Rad recommends the use of eight-channel manual P-50 Pipetman with 200 μL tips (not wide or narrow bore), and not the P-20 or P-1000 Pipetman. With the cartridge holder being flat, they suggest aspirating 40 μL of the droplets (not more) with the pipet at a 30–45° angle from the vertical direction at the junction of the side wall and the bottom. The droplets should be dispensed near but not at the bottom of the wells of the PCR plate slowly.

References

1. Manoj P (2014) Droplet digital PCR technology promises new applications and research areas. Mitochondrial DNA. Apr 29. [Epub ahead of print] doi:10.3109/19401736.2014.913168
2. Brunstein J (2013) Digital PCR: theory and applications. MLO Med Lab Obs 45:34–35
3. Hindson BJ, Ness KD, Masquelier DA et al (2011) High-throughput droplet digital PCR system for absolute quantitation of DNA copy number. Anal Chem 83:8604–8610
4. Huggett JF, Foy CA, Benes V et al (2013) The digital MIQE guidelines: minimum Information for publication of quantitative digital PCR experiments. Clin Chem 59:892–902
5. Taly V, Pekin D, Benhaim L et al (2013) Multiplex picodroplet digital PCR to detect KRAS mutations in circulating DNA from the plasma of colorectal cancer patients. Clin Chem 59:1722–1731
6. Nakano M, Komatsu J, Matsuura S et al (2003) Single-molecule PCR using water-in-oil emulsion. J Biotechnol 102:117–124
7. Vogelstein B, Kinzler KW (1999) Digital PCR. Proc Natl Acad Sci U S A 96: 9236–9241
8. Azuara D, Ginesta MM, Gausachs M et al (2012) Nanofluidic digital PCR for KRAS mutation detection and quantification in gastrointestinal cancer. Clin Chem 58: 1332–1341
9. Whale AS, Huggett JF, Cowen S et al (2012) Comparison of microfluidic digital PCR and conventional quantitative PCR for measuring copy number variation. Nucleic Acids Res 40:e82
10. Bizouarn F (2014) Introduction to digital PCR. Methods Mol Biol 1160:27–41
11. Gevensleben H, Garcia-Murillas I, Graeser MK et al (2013) Noninvasive detection of HER2 amplification with plasma DNA digital PCR. Clin Cancer Res 19:3276–3284
12. Garcia-Murillas I, Lambros M, Turner NC (2013) Determination of HER2 amplification status on tumour DNA by digital PCR. PLoS One 8:e83409
13. Weaver S, Dube S, Mir A et al (2010) Taking qPCR to a higher level: analysis of CNV reveals the power of high throughput qPCR to enhance quantitative resolution. Methods 50:271–276
14. Belgrader P, Tanner SC, Regan JF et al (2013) Droplet digital PCR measurement of HER2 copy number alteration in formalin-fixed paraffin-embedded breast carcinoma tissue. Clin Chem 59:991–994
15. Heredia NJ, Belgrader P, Wang S et al (2013) Droplet digital PCR quantitation of HER2 expression in FFPE breast cancer samples. Methods 59:S20–S23
16. Wang J, Ramakrishnan R, Tang Z et al (2010) Quantifying EGFR alterations in the lung cancer genome with nanofluidic digital PCR arrays. Clin Chem 56:623–632
17. Hindson CM, Chevillet JR, Briggs HA et al (2013) Absolute quantification by droplet digital PCR versus analog real-time PCR. Nat Methods 10:1003–1005
18. Jennings LJ, George D, Czech J et al (2014) Detection and quantification of BCR-ABL1 fusion transcripts by droplet digital PCR. J Mol Diagn 16:174–179
19. Sanders R, Mason DJ, Foy CA et al (2013) Evaluation of digital PCR for absolute RNA quantification. PLoS One 8:e75296
20. Kinz E, Leiherer A, Lang AH et al (2014) Accurate quantitation of JAK2 V617F allele burden by array-based digital PCR. Int J Lab Hematol 37(2):217–24
21. Kelley K, Cosman A, Belgrader P et al (2013) Detection of methicillin-resistant Staphylococcus

aureus by a duplex droplet digital PCR assay. J Clin Microbiol 51:2033–2039
22. Henrich TJ, Gallien S, Li JZ et al (2012) Low-level detection and quantitation of cellular HIV-1 DNA and 2-LTR circles using droplet digital PCR. J Virol Methods 186:68–72
23. Strain MC, Lada SM, Luong T et al (2013) Highly precise measurement of HIV DNA by droplet digital PCR. PLoS One 8:e55943
24. Hatano H, Strain MC, Scherzer R et al (2013) Increase in 2–long terminal repeat circles and decrease in D-dimer after raltegravir intensification in patients with treated HIV infection: a randomized, placebo-controlled trial. J Infect Dis 208:1436–1442
25. Lench N, Barrett A, Fielding S et al (2013) The clinical implementation of non-invasive prenatal diagnosis for single-gene disorders: challenges and progress made. Prenat Diagn 33:555–562
26. Barrett AN, Chitty LS (2014) Developing noninvasive diagnosis for single-gene disorders: the role of digital PCR. Methods Mol Biol 1160:215–228

Chapter 6

Quantitative PCR for Plasma Epstein-Barr Virus Loads in Cancer Diagnostics

Sanam Loghavi

Abstract

Epstein-Barr virus (EBV) is the causative agent of infectious mononucleosis and is associated with post-transplant lymphoproliferative disease (PTLD), Hodgkin's lymphoma, Burkitt's lymphoma, nasopharyngeal carcinoma, and HIV-related lymphomas. It infects nearly all humans and then persists for the life of the host in a small proportion of benign B lymphocytes. EBV reactivation, usually in the setting of immunosuppression, is the main risk factor for development of EBV-associated malignancies. EBV reactivation can be detected in tissue specimens using EBV-encoded RNA (EBER) in situ hybridization (ISH), which is routinely used for diagnosis of PTLD and nasopharyngeal carcinoma. However, EBER ISH cannot be routinely used for screening asymptomatic or monitoring posttreatment outcome due to difficulty in obtaining tissue specimens for testing and the nonquantitative nature of the assay. Recent studies have shown that EBV genomic DNA can be detected in blood of patients with EBV-associated diseases, and that monitoring of EBV viral load in blood could be an effective method of distinguishing disease-associated EBV reactivation from incidental EBV present in benign B lymphocytes, and could be used for diagnostic screening and monitoring of EBV-associated diseases. In this chapter we discuss a protocol for quantitative plasma EBV DNA analysis.

Key words EBV, PTLD, Nasopharyngeal carcinoma, qPCR

1 Introduction

Epstein-Barr virus (EBV), the causative agent of infectious mononucleosis, is implicated in a number of malignancies as an etiologic factor including nasopharyngeal carcinoma (NPC). Other tumor types in which EBV plays an etiologic role include certain subtypes of Hodgkin lymphoma, Burkitt lymphoma, T/NK cell lymphomas, human immunodeficiency virus (HIV)-related lymphomas, and posttransplant lymphoproliferative disorders [1, 2]. Monitoring plasma/serum EBV viral loads using polymerase chain reaction (PCR), initially introduced for this purpose in the late 1990s [3, 4], has proven to be effective in diagnosis, prognostication, and disease monitoring in patients with NPC and other EBV-derived malignancies. Previous studies have shown that EBV DNA is present

Rajyalakshmi Luthra et al. (eds.), *Clinical Applications of PCR*, Methods in Molecular Biology, vol. 1392,
DOI 10.1007/978-1-4939-3360-0_6, © Springer Science+Business Media New York 2016

in the serum of patients with NPC and usually absent or present at very low levels in healthy individuals, indicating the relatively high specificity of this test as a diagnostic tool [5–7].

The protocol here describes a quantitative PCR (qPCR) assay using the ABI Prism 7900HT Sequence Detection System (Perkin-Elmer, Applied Biosystems) for estimating EBV load in plasma samples. The approach uses dual labeled fluorogenic hybridization TaqMan probes, where one fluorescent dye serves as a reporter (FAM, i.e., 6-carboxyfluorescein) and a second fluorescent dye, TAMRA (i.e., 6 carboxy-tetramethyl-rhodamine), serves as quencher. The exonuclease activity of Taq polymerase cleaves the probe and separates the reporter dye from the quencher dye, resulting in an increase in the fluorescent reporter signal. A spiked control sequence is used to control for the efficacy of extraction and amplification. This method can be used in clinical situations for detection of EBV including in immunosuppressed patients at risk for EBV-driven lymphoproliferative disorders. Testing can assist in initial diagnosis or in monitoring the efficacy of therapeutic regimens.

2 Materials

2.1 Sample Requirement

Whole blood (or bone marrow) collected in ethylenediaminetetraacetic acid (EDTA) containing collection tubes.

2.2 DNA Extraction

1. QIAsymphony ® SP instrument.
2. QIAsymphony ® DSP Virus/Pathogen Mini Kit.
   (a) Reagent Cartridge.
   (b) Enzyme Rack.
   (c) Piercing Lid.
   (d) Buffer AVE.
   (e) Carrier RNA.
   (f) Reuse Seal Set.
3. TaqMan ® Exogenous Internal Positive Control Reagents (VIC ® Probe) (Life Technologies).
4. 8-Well Sample Prep Cartridges (QIAgen ®).
5. 8-Rod Covers (QIAgen ®).
6. Filter Tips, 1500 μl, Qsym ® SP (QIAgen ®).
7. Filter Tips, 200 μl (QIAgen®).
8. Accessory Troughs (QIAgen®).
9. Elution Microtubes CL (QIAgen®).
10. 2.0 ml tubes with screw on caps.
11. 1.5 ml tubes with snap caps.

2.3 Real-Time PCR Assay

1. ABI Prism 7900HT Sequence Detection System (Applied Biosystems).
2. ABI MicroAmp® Optical 96-well Reaction Plate (Applied Biosystems).
3. MicroAmp® Optical Caps (Applied Biosystems).
4. TaqMan® 2× Universal PCR Master Mix, no AmpErase® UNG (Life Technologies).
   (a) AmpliTaq Gold® DNA Polymerase, UP (Ultra Pure).
   (b) dNTPs with dUTP.
   (c) ROX™ Passive Reference.
   (d) Optimized buffer components.
5. TaqMan Exogenous Internal Positive Control Reagents (Life Technologies).
   (a) 10× Exo IPC Mix.
   (b) 10× Exo IPC Block.
   (c) 50× Exo IPC DNA.
6. Namalwa Burkitt lymphoma cell line.
7. tRNA.
8. Sterilized Type I water.
9. Primers (*see* **Note 1**).
   (a) EBV Forward Primer: EBV W-F 5′-GCA GCC GCC CAG TCT CT-3′.
   (b) EBV Reverse Primer: EBV W-R 5′-ACA GAC AGT GCA CAG GAG CCT-3′.
10. Probe: EBVBamH1W 5′-6FAM-AAA AGC TGG CGC CCT TGC CTG TAMRA.

3 Methods

3.1 Sample Collection and Preparation

1. Collect whole blood into EDTA containing blood collection tube. A minimum of 300 μl of whole blood is required (*see* **Note 2**).
2. Centrifuge the sample at 1500 rpm (453 rcf/g-force) for 10 min to separate plasma from cellular components.
3. Transfer the supernatant into a 2 ml tube with screw on cap. Do not take any of the buffy coat or cellular components when removing the supernatant.

3.2 DNA Extraction

DNA extraction is performed following the "blood and body fluid protocol" as per manufacturer's instructions. We use the QIAsymphony DNA extraction system for whole blood specimens [8].

1. Internal Control Preparation:
   (a) Use 1.8 μl of internal control per sample.
   (b) Transfer the appropriate amount of internal control and Buffer AVE into a 2.0 ml tube with screw on cap tube. Replace Carrier RNA with buffer AVE.
2. QIAsymphony ® SP instrument Preparation:
   (a) Close all drawers and the hood.
   (b) Switch on the QIAsymphony ® SP instrument, and wait until the "Sample Preparation" screen appears and the initialization procedure has finished.
   (c) Log in to the instrument.
   (d) Ensure the "Waste" drawer is prepared properly, and perform an inventory scan of the "Waste" drawer, including the tip chute and liquid waste. Replace the tip disposal bag if necessary.
   (e) Load the microwell plate with the plate adapter on the first position (slot 1) in the "Eluate" drawer.
   (f) The instrument will be showing the eluate plate screen. Select the eluate rack slot that will be used (slot 1).
   (g) Under the Configure Tab > "Available Rack Types" choose "deep well," then arrow down to 19588 EMTR.
   (h) Enter a number in the reserve column section of the screen if there is a column of tubes used/missing on the EMTR plate. Otherwise leave as zero.
   (i) Select "Edit ID" and name your plate.
   (j) Select OK. The instrument will now scan the plate.
3. Reagent Cartridge Preparation:
   (a) If using an unopened reagent cartridge, carefully position the piercing lid over the cartridge and snap ends into place. The lid must be installed with the rounded edge at the free end of the cartridge (i.e., not over the magnetic bead trough). Also be sure that each piercing column is centered directly over a cartridge hole, or cartridge may not pierce completely leading to run failure. Label the cartridge with the date and tech initials.
   (b) Remove the magnetic bead trough and vortex for 3 min.
   (c) Uncap the Proteinase K enzyme rack tubes and place caps on the storage shelf (on side of grey reagent rack).
   (d) Slide the enzyme rack onto the side of the reagent rack.
   (e) Load the rack containing the reagents on the QIAsymphony.
   (f) Load the required reagent cartridge(s) and consumables into the "Reagents and Consumables" drawer. For 24

samples (1 batch), there should be 21 sample prep cartridges (1 box), 8-rod covers (partial box), 1500 μl tips (2½ racks), 200 μl tips (~1 rack) (*see* **Note 3**).

(g) Press down on all cartridges to ensure they are seated firmly in the rack.

4. Loading EBV Samples:

(a) Load the Sample Drawer. First place 2.0 ml tube inserts (3b size) into the 24 tube rack. Place inserts only in the positions where samples will be located. If there are empty spots on the rack with inserts in them, the robot will look for samples at those locations.

(b) Add the samples to the rack, removing the caps.

(c) Open the sample drawer and line up your tube rack containing samples with the black line on the drawer lid.

(d) The sensor will activate, and as it does the green arrow will start blinking. As it blinks, insert the rack all the way in. The arrows will turn orange if it was inserted correctly. If the arrows are red, an error appears on the screen—click OK and repeat insertion procedure.

(e) On the screen, select the "Batch" icon corresponding to the first rack loaded (e.g., Batch 1 for rack in Lane 1). The samples will appear on the next screen for naming and protocol selection.

(f) After labeling the samples, click "Next," then "Select All" and choose the "Viral" Tab. Under the Viral tab select cell free 200 v5 EBV DNA extraction.

(g) Select "Next," the instrument will go to the eluate screen. Choose slot to which samples will be eluted (e.g., slot 1), and then choose the elution volume of 60 μl.

(h) Select "View Rack" to check the plate configuration. Select "Queue."

(i) The screen goes back to Sample View. If running additional batches select each batch one at a time and follow **steps e–g** to add to the eluate plate. When finished, select RUN on the Sample View screen.

(j) The instrument will ask to do a reagent scan before running. Choose "Yes."

5. After scanning, the instrument will either proceed with the run or alert user to missing items with a pop-up message. Hit OK on the message box. The reagents and consumables list will appear with missing items highlighted in pink. A new reagent cartridge will be needed if any reagents are low. After loading more items, choose "Scan Reagents" and select from the list only the items which were replenished to save time.

6. Loading Internal Control:
   (a) Load the internal control on the sample rack similarly to the samples. Insert the rack with the internal control in slot "A" designated for the internal control.
   (b) Select the tab "IC" on the screen and attach the same protocol cell free 200 v5 to the internal control tube(s).
7. Choose "Run" to start the instrument. Each full rack of 24 samples will take 1 h 20 min to complete.
8. When run is complete, instrument LED lights will flash and all batches will show 100 % completion. Select the "E" button to go to the eluate plate screen. Choose the eluate plate slot (slot 1) and then the "Configure" tab, and then open the eluate drawer. Hit the "remove plate" button. Choose "Yes" when asked if the plate should be removed. Take the eluate plate and holder out of the drawer and cover the sample wells with QIAgen cap covers.
9. Close the eluate drawer and choose OK on the screen. Drawer will be scanned to confirm plate removal and clear out plate configuration. While scanning, remove the reagent cartridge and cover wells with cap strips to avoid evaporation.
10. Remove sample racks from the sample drawer. Match labeled 1.5 ml tubes to the sample input tubes to ensure correct order.
11. Carefully transfer EBV samples one at a time from eluate plate to the 1.5 ml tubes and place in the 4 °C refrigerator. Discard the input sample tubes in a biohazard waste container.
12. Open the waste drawer and empty the liquid waste bottle. Throw away any full consumables boxes and replace with empty ones from the reagents/consumables drawer. Empty the ethanol trough and invert on paper towel to dry. Replace through weekly.
13. Close all drawer and logout of instrument. Power off.

3.3 Positive and Negative Controls

1. Preparation of EBV Standards from Namalwa Cell Line.
   (a) EBV DNA extracted from the Namalwa cell line (as per instructions above) is used to establish a calibration curve. Quantification of circulating EBV DNA is performed based upon this calibration curve.
   (b) After obtaining a homogeneous DNA solution, OD 260 is measured by spectrophotometry and the DNA concentration is calculated. Namalwa cell DNA contains $\times 10^5$ copies EBV/μg DNA (EBV copies/μl DNA = Namalwa cell line DNA concentration $\times (3 \times 10^5)$) [9] (*see* **Note 4**).
   (c) The newly extracted Namalwa cell line DNA is considered the stock (undiluted) concentration. Make a five-point,

tenfold serial dilution of the stock DNA (stock, 10^{-1}, 10^{-2}, 10^{-3}, 10^{-4}, and 10^{-5}) in DNA hydration buffer in order to generate a standard curve.

(d) In order to generate a new average slope and intercept for the stock, the standards prepared from this stock are run together with the current standard in multiple runs (5 times in four different runs) to obtain a total of 20 different run points.

(e) Compile all run point data. Calculate the new slope and intercept ranges according to the following formula:

- $Y = -MX + C$ where $-M$ = slope (M = slope; C = intercept; Y = CT; X = log quantity).

2. Preparation of high and low concentration positive controls (Namalwa DNA) (*see* **Note 5**).

(a) High and low positive controls (HPC and LPC, respectively) are prepared from Namalwa cell line. The cell line is diluted in a DNA Suspension Buffer (DSB) and tRNA as shown below:

- Stock (107).
- 10–1: 30 μl stock + 270 μl DSB.
- 10–2: 100 μl 10–1 + 900 μl DSB.
- 10–3: 30 μl 10–2 + 270 μl DSB.
- 10–4: 100 μl 10–3 + 900 μl DSB.
- Dilute 10^{-2} and 10^{-4} with equal proportions (1:1) of yeast tRNA to obtain high and low positive controls, respectively. (Yeast tRNA is reconstituted with DSB to a concentration of 0.05 μg/μl.)
  - Preparation of tRNA from dry powder

    Reconstitute powder tRNA in DSB to achieve a concentration of 10 μg/μl (*see* **Note 6**).

    Reconstitute the above product to achieve a final concentration of 0.05 μg/μl. This concentration will be used as your diluent.

(b) We established the mean copy number for HPC and LPC by analyzing them in 20 runs (4× in 5 assays). These controls are to be included in each run, in duplicates, and the copy number obtained for each level should be recorded. The values should fall within two standard deviations of the established mean for each level as shown in Table 1.

3. A negative control (sterile type I water) should be included with every run.

4. A no amplification control (NAC) (diluted internal positive control + an internal positive control blocking agent) is included in every run.

Table 1
Reference values for high and low concentration positive controls prepared from Namalwa cell line

	HPC-a mean EBV DNA copy number	LPC-a mean EBV DNA copy number
	141868.6	941.5
	HPC-a range	LPC-a range
2 SD	102323.8–181413.4	346.0–1537.1
3 SD	82551.4–201185.9	48.2–1834.9

HPC high positive control, *LPC* low positive control, *SD* standard deviation

EBV qPCR MASTER MIX

		Final Concentration	EBV	
			µl/rxn	1 rxns
	2X Universal Mix *with UNG*	1x	15.00	15.00
Primers	EBV BamH1W Forward 20 µM	0.4µM	0.60	0.60
	EBV BamH1W Reverse 20 µM	0.4µM	0.60	0.60
	H_2O		0.50	0.50
Probe	EBV BamH1W FAM PROBE 20 µM	0.2 µM	0.30	0.30
	IPC primer-probe mix (VIC)	1X	3.00	3.00
		Total	20.00	20.00
		Sample*	10	

STD1-5: 1:10 serial dilution starting from Stock (5µl DNA + 45µl DNA suspension buffer)
Primers: Add 32µL of water to aliquoted 100µM primer to make 20µM working primer.
Probe: Add 16µL of water to aliquoted 100µM probe to make 20µM working probe. (50 rxns)

** Diluted IPC: 4ml IPC DNA + 96ml H2O

Number of Standards and Positive Ctrls	µL/rxn	25
IPC DNA (µL)	0.2	5.0
H2O (µL)	4.8	120.0
Total amount of diluted IPC (µL)	5	125.0

Reagents
2x Universal Mix
EBV BamH1W-F
EBV BamH1W-R
H_2O
EBV BamH1W-Probe
IPC Primer-Probe mix
50x EXO IPC DNA
DNA Suspension Buffer

Fig. 1 Epstein-Barr virus qPCR master mix worksheet. This worksheet is an example of one used at our laboratory for preparation of the master mix and internal positive control for this assay. *EBV* Epstein-Barr virus, *IPC* internal positive control, *µl* microliter, *µM* micromolar, *rxn* reactions, *UNG* Uracil-N glycosylase, VIC: *EXO* exogenous

3.4 Real-Time PCR Assay

1. Make a 1:25 dilution of the internal positive control (IPC) by adding 4 µl of IPC to 96 µl of Type I water (*see* Fig. 1).
2. Prepare the master mix according to the worksheet example provided for the desired number of reactions (*see* Fig. 1).
3. Add 20 µl of master mix to the appropriate wells of a 96-well optical PCR plate
4. Add the controls and water to the appropriate wells in the following manner:
   (a) NAC—Add 5 µl of Exo IPC Block to each well + 5 µl IPC dilution

Table 2
Suggested thermal profile for ABI prism 7900HT sequence detector

50 °C 95 °C	2 min 10 min	
95 °C 60 °C	15 s 1 min	×40 Cycles

(b) No template control (NTC)—Add 5 μl H2O +5 μl of IPC dilution

(c) STDs—Add 5 μl of standard DNA + 5 μl IPC dilution

(d) HPC—Add 5 μl of HPC + 5 μl IPC dilution.

(e) Add 5 μl of LPC + 5 μl IPC dilution.

5. Add 10 μl of patient DNA in duplicates to the appropriate wells.
6. Cover each row on the plate with an 8-well optical cap strip.
7. Vortex briefly to mix and remove any bubbles in the wells.
8. Centrifuge briefly at no more than 2000 rpm (805 rcf/g-force) and ensure that there are no bubbles.
9. Place the 96-well plate inside the reaction chamber of the ABI Prism 7900HT Sequence Detector (*see* Table 2 for suggested thermal profile).
10. Input the appropriate sample designations into the sequence detector program software.

3.5 EBV DNA Transcript Quantitation

1. EBV DNA quantity is determined in this assay by running a set of standards of known quantity and measuring their cycle threshold (CT). A graph is generated using this data with Log 10 concentration/quantity of the standards on the X-axis and CT values on the Y-axis. A line that best fits is drawn through the data points and the slope and intercept calculated. ($X = C - Y/M$: M = slope; C = intercept; Y = CT; X = log quantity) (*see* **Notes** 7 and **8**).

4 Notes

1. To preserve the integrity of the assay and to prevent disintegration of primers and probe from repeated freezing and thawing, aliquots of the 100 μM forward and reverse primers and probe can be made. The 8 μl aliquots of 100 μM primer stock can be reconstituted to a working concentration of 20 μlM by adding 32 μl of molecular grade water. The 4 μl aliquots of 100 μM probe can be reconstituted to a working solution of 20 μM by adding 16 μl of water.

2. Specimens can be refrigerated for up to 48 h prior to delivery to the lab. Frozen plasma may also be submitted (in the separation of plasma from whole blood, ensure that you do not take any of the buffy coat or cellular elements when removing the plasma for testing).
3. Loading only completely filled tip racks will help the instrument scan more efficiently. Consolidate partially used racks as needed.
4. Each new extraction of Namalwa cell DNA has to be quality controlled and calibrated. The new stock concentration is run with a current standard to determine CT and copy number. If the CT of the new stock is less than the CT of the current stock, then the new stock must be diluted and rerun. If the CT of the new stock is greater than the CT of current stock, then the new stock must be re-precipitated and rehydrated in a smaller volume of diluent than before and rerun. When <1CT difference between the new and current stock is obtained, the new stock is ready to be tested. Newly prepared stock vials of Namalwa DNA are fairly viscous. Place the stock vial at 30 ° C for 30 to 60 min prior to proceeding with your standard preparation. Stock Namalwa cell line DNA is stored at −20 °C in 50 μl aliquots. Serially diluted standards are freshly prepared from aliquots of stock DNA each time they are to be included in a run.
5. Store Controls in −80 °C.
6. Store tRNA at −80 °C.
7. The lower limit of linearity of this assay is 400 EBV copies/ml plasma. EBV detected below this level is reported as "low positive."
8. Troubleshooting:
   (a) Rules for high and low positive controls used to monitor ongoing test performance and to institute corrective action:
      - 1 3S: any value > +/-3SD; reject run; repeat run.
      - Both high and low controls > +/-2SD; reject run; repeat run.
      - 10×: 10 values are on same side of the mean (indicates a trend/shift): review; investigate cause and perform corrective action.
      - If IPC (internal positive control) fails, repeat DNA extraction.
      - If duplicate samples vary by >2 CT, repeat PCR.
      - If NTC (no template control) is positive for EBV, reject.
      - If NAC (no amplification control) is positive for IPC, reject.

References

1. Gulley ML, Tang W (2010) Using Epstein-Barr viral load assays to diagnose, monitor, and prevent posttransplant lymphoproliferative disorder. Clin Microbiol Rev 23:350–366. doi:10.1128/cmr.00006-09
2. Gulley ML (2001) Molecular diagnosis of Epstein-Barr virus-related diseases. J Mol Diagn 3:1–10. doi:10.1016/s1525-1578(10)60642-3
3. Lo YM, Chan LY, Lo KW et al (1999) Quantitative analysis of cell-free Epstein-Barr virus DNA in plasma of patients with nasopharyngeal carcinoma. Cancer Res 59:1188–1191
4. Mutirangura A, Pornthanakasem W, Theamboonlers A et al (1998) Epstein-Barr viral DNA in serum of patients with nasopharyngeal carcinoma. Clin Cancer Res 4:665–669
5. Ryan JL, Fan H, Glaser SL et al (2004) Epstein-Barr virus quantitation by real-time PCR targeting multiple gene segments: a novel approach to screen for the virus in paraffin-embedded tissue and plasma. J Mol Diagn 6:378–385. doi:10.1016/s1525-1578(10)60535-1
6. Fan H, Robetorye RS (2013) Epstein-Barr virus (EBV) load determination using real-time quantitative polymerase chain reaction. Methods Mol Biol 999:231–243. doi:10.1007/978-1-62703-357-2_17
7. Yip TT, Ngan RK, Fong AH et al (2014) Application of circulating plasma/serum EBV DNA in the clinical management of nasopharyngeal carcinoma. Oral Oncol 50:527–538. doi:10.1016/j.oraloncology.2013.12.011
8. Laus S, Kingsley LA, Green M et al (2011) Comparison of QIAsymphony automated and QIAamp manual DNA extraction systems for measuring Epstein-Barr virus DNA load in whole blood using real-time PCR. J Mol Diagn 13:695–700. doi:10.1016/j.jmoldx.2011.07.006
9. Klein G, Dombos L, Gothoskar B (1972) Sensitivity of Epstein-Barr virus (EBV) producer and non-producer human lymphoblastoid cell lines to superinfection with EB-virus. Int J Cancer 10:44–57

Chapter 7

High-Resolution Melt Curve Analysis in Cancer Mutation Screen

Meenakshi Mehrotra and Keyur P. Patel

Abstract

High-resolution melt (HRM) curve analysis is a PCR-based assay that identifies sequence alterations based on subtle variations in the melting curves of mutated versus wild-type DNA sequences. HRM analysis is a high-throughput, sensitive, and efficient alternative to Sanger sequencing and is used to assess for mutations in clinically important genes involved in cancer diagnosis. The technique involves PCR amplification of a target sequence in the presence of a fluorescent double-stranded DNA (dsDNA) binding dye, melting of the fluorescent amplicons, and subsequent interpretation of melt curve profiles.

Key words High-resolution melt curve analysis, PCR, Double-stranded DNA binding fluorescent dyes

1 Introduction

High-resolution DNA melting (HRM or HRMA), a post-PCR method, appeared as a rapid method for genotyping known variants or scanning unknown variants by measuring changes in the melting of a DNA duplex [1–3]. The method is highly sensitive and can discriminate DNA sequence variants based on insertions or deletions, single nucleotide polymorphisms, and homo- or heterozygosity [4, 5]. Melting of double-stranded DNA molecules is influenced by several factors, including DNA length, GC content, and sequence. Duplex melting is generally monitored using intercalating dyes, which bind to double-stranded DNA. A homozygous sequence variant usually changes the Tm of the duplex. Recent advances in DNA melting techniques and development of new fluorescent DNA binding dyes allow for more precise assessment of sequence variations based on melting analysis, without the need for labeled probes, and have the potential to greatly decrease the burden of sequencing [6]. High-resolution melting analysis involves three main steps: (1) PCR amplification of a target region to high copy numbers in the presence of a fluorescent double-stranded (ds) DNA binding dye, (2) melting of the fluorescently

Rajyalakshmi Luthra et al. (eds.), *Clinical Applications of PCR*, Methods in Molecular Biology, vol. 1392,
DOI 10.1007/978-1-4939-3360-0_7,

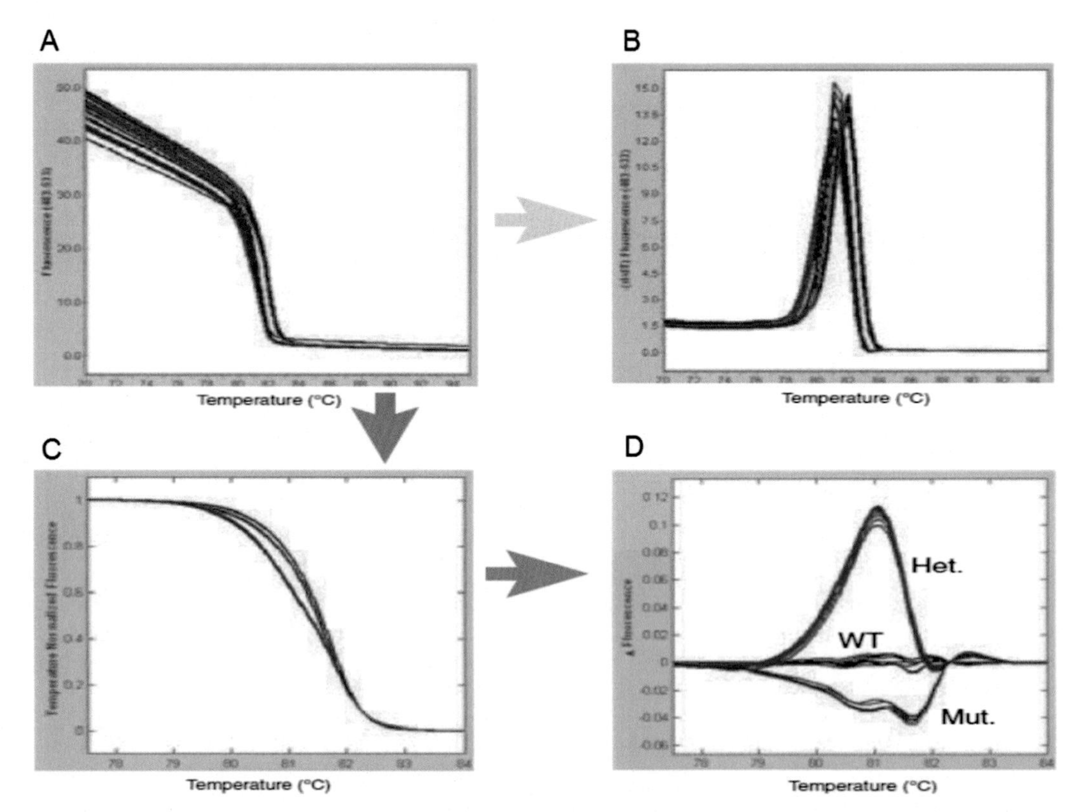

Fig. 1 High-resolution melting analysis: (**a**) Samples analyzed in presence of an intercalating dye decrease the fluorescence when the double-stranded amplicon denatures. (**b**) Melting point differences for whole amplicons are often too low for unambiguous differentiation of the various genotypes in the negative first-derivative plot. (**c**) Better separation of melting curve shapes after signal normalization (Heterozygote PCR products, forming hetero-duplexes, begin to denature at lower temperatures than homozygotes). (**d**) Differences plot plotted for fluorescence and temperature variation as compared to wild type leading to clear separation of sample

labeled product, and (3) analysis of melt curves. A heterozygous sample contains four duplex species and its observed melting curve is a composite of the four individual melting curves. Exchange between G: C and T: A base pairs makes the large change in Tm: approx. 0.8–1.4 °C whereas if the bases swap strands the change in Tm is smaller. HRMA is a simple method, after PCR, carried out in the presence of a suitable dye; the product is heated while the level of fluorescence is measured. As the temperature rises and the duplex passes through its melting transition, dye is released and fluorescence intensity is reduced. For mutation scanning, a group of raw melting curves show variation in both the fluorescence and temperature axes (Fig. 1a). Normalization of fluorescence before and after the melting transition corrects fluorescence variation caused by factors such as variable amplicon levels whereas normalization of temperature, known as temperature shifting, is a necessary correction for these variations, but at the expense of homozygote detection (Fig. 1b, c). Melting curves are often converted into a subtractive difference plot in which the sample melting curves are shown relative to a wild-type control curve. Difference plots are more informative than derivative plots, and can make even subtle changes in the shape of the melting curve readily apparent (Fig. 1d) [4, 5, 7, 8].

2 Materials

2.1 Equipment

1. Microcentrifuge.
2. Vortex.
3. Pipets for PCR.
4. Filter tips.
5. Millipore water.
6. 1.5-mL microcentrifuge tubes.
7. Roche light cycler 480.
8. LightCycler 480 Sealing Foil.
9. LightCycler 480 Multi-well Plate 96.
10. Roche analysis software version 1.5.0 SP3.
11. Foil Sealer.
12. ABI 3130 and 3730 with associated software for sequencing analysis.
13. Optical adhesive film.
14. Optical 96-well plate (ABI).
15. Optical caps.
16. CLP 3410 Centrifuge.
17. Qiagen 4-15C Centrifuge.
18. ABI 2720 Thermal cycler.
19. Qiaxcel and related software.
20. Cryo-rack.
21. Latex Gloves.

2.2 Reagents

1. Molecular biology grade water (DNase free, RNase free, Protease free)-ATCC.
2. Genomic DNA samples.
3. LightCycler 480 high resolution melting master mix (Roche) containing Ampli*Taq* Gold DNA polymerase and dNTPs.
4. 25 mM $MgCl_2$.
5. Positive and negative control cell lines (HL60 cell line used as a wild-type control).

6. M13-Tagged *KIT* forward and reverse PCR primer Mix (5 μM).
7. M13 sequencing primers (1 μM).
8. POP 7 polymer for electrophoresis (ABI).
9. 3730 buffer (10×) with EDTA (ABI).
10. BigDye terminator v1.1 ready reaction cycle sequencing kit (ABI).

2.3 Primer Sequences for KIT Exon 17

M13-*KIT*-FORWARD PRIMER.

5′-TGT AAA ACG ACG GCC AGT ATG GTT TTC TTT TCT CCT CC-3′.

M13-*KIT*-REVERSE PRIMER.

5′-CAG GAA ACA GCT ATG ACC TGC AGG ACT GTC AAG CAG AG-3′.

3 Methods

1. DNA was extracted from bone marrow aspirate specimens using the Autopure extractor (QIAGEN/Gentra, Valencia, CA).
2. All HRM PCR amplification analyses were performed in duplicate.
3. PCR was performed in a 96-well plate with a 20-μL volume including 50 ng DNA, 5 μm of each primer, 2.5 mmol/L of $MgCl_2$, 4.7 μL of water, and 10 μL of LightCycler 480 high resolution melting master mix (Roche, Indianapolis, IN). The reaction mix was subjected to initial denaturation at 95 °C for 10 min, followed by 45 cycles of amplification consisting of denaturation at 95 °C for 30 s, annealing at 58 °C for 30 s, and extension at 72 °C for 12 s. Melting was performed with a denaturation step at 95 °C for 1 min, followed by annealing at 40 °C for 1 min and a melt from 70 to 97 °C at a ramp rate of 0.03 °C/s with 18 acquisitions per second. LightCycler 480 resolight dye (Roche), a fluorescent dye that uniformly binds to the minor groove of double-stranded DNA in a non-sequence-dependent manner, was used to detect mutations/sequence variations in the entire length of *KIT* exon 17 amplicons.
4. The Roche analysis software version 1.5.0.SP3 (Roche) was used to detect variant sequences, and the settings were optimized to achieve maximum sensitivity.
5. For *KIT exon 17*, the parameters selected were pre-melt window at 79.53–80.34 °C, post-melt window at 83.05–83.66 °C, range of Cp values from 23 to 27, Tm shift threshold at 3.226, and sensitivity at 0.4.
6. To validate results of *KIT* mutations detected in Exon 17 through HRM analysis, sequencing of amplicons was performed

by Sanger sequencing using the forward and reverse primers tagged with M13 universal sequences: M13 forward, 5′-TGTAAAACGACGGCCAGT-3′ or M13 reverse, 5′-CAGGAAACAGCTATGACC-3′.

7. After HRM analysis, PCR products were purified using AMPure magnetic beads (Agentcourt, Danvers, MA) according to the manufacturer's protocol. Purified amplicons were diluted (1:5) in water and 5 μL was used for Sanger sequencing using 3730 DNA Analyzer (Applied Biosystems, Carlsbad, CA).
8. The resulting data were analyzed by SeqScape software versions 2.5 and/or 2.7 (Applied Biosystems).

4 Results

1. Most common mutation for *KIT* in codon D816 was found in approximately 30–40 % of AML patients with inv(16) which is associated with significantly higher incidence of relapse, as well as lower survival [9].
2. The HRM assays were established using positive (KASUMI) and negative (HL60) cell line for *KIT* exon 17 mutation.
3. The differential melting of the variant sequences compared to the wild-type sequences was detected by plotting increase in temperature on the x axis versus the normalized fluorescence values or the rate of change in fluorescence per unit of change in temperature on the y axis to create a normalized melting curve or a derivative plot.
4. The difference in the fluorescence of test sequence versus wild-type sequence was plotted to create a normalized and temperature-shifted difference plot.
5. For wild-type *KIT*, a single melting peak was noted at Tm 81.6 °C. A difference plot analysis using a pre-melt temperature range of 79.53–80.34 °C and post-melt temperature range of 83.05–83.06 °C detected all possible variant sequences.
6. The Cp values for *KIT* HRMA should be in range of 23–27.
7. 127 Patient samples screened for *KIT* exon 17 mutation detection by HRM analysis.
8. Out of 127 samples 35.4 % ($n=45$) was found to be positive, 51.1 % ($n=65$) negative, and 13.3 % ($n=17$) showed discordance for KIT exon 17 mutations.
9. 48.8 % ($n=62$) positive samples were subjected to Sanger sequencing for further confirmation. Out of 62 samples submitted for confirmation 52 were positive and 10 were negative for *KIT* exon 17 mutation.

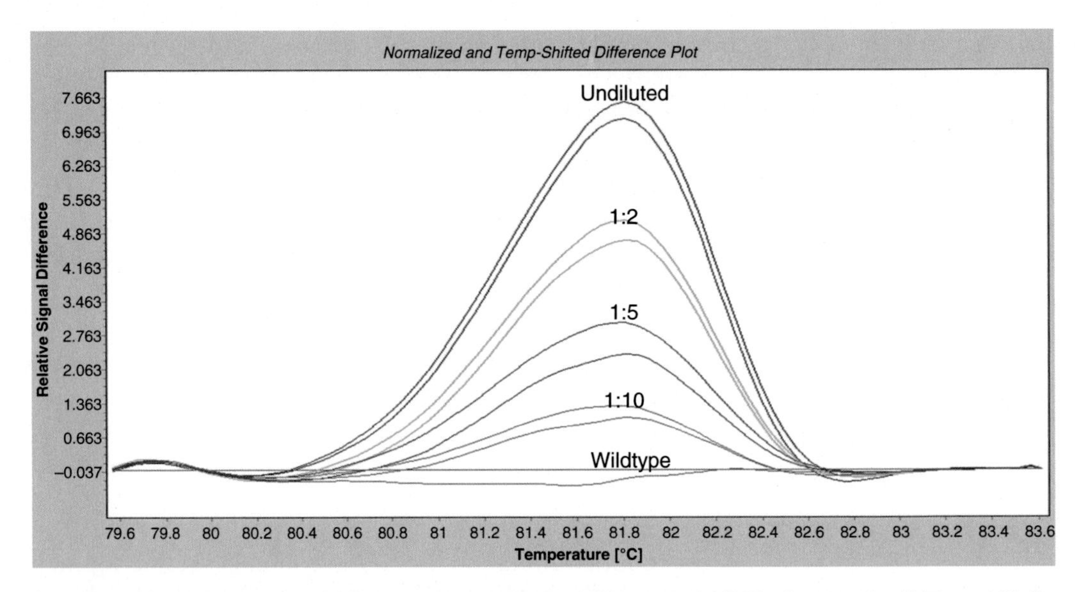

Fig. 2 Sensitivity of detection of *KIT* exon 17 mutation by HRM analysis. Ratio denotes the % Tumor dilution (1:2 = 50 %; 1:5 = 25 %; 1:10 = 10 %)

10. Sensitivity for detection of *KIT* Exon 17 mutation by HRM was performed by diluting KSAUMI with HL60 at 50, 25, and 10 %.
11. Limit of detection for HRM assay is 10 % as compared to 5 % for Sanger sequencing (Fig. 2).

5 Notes

1. Cp values >30 indicate a problem with the PCR setup or amplification (Subheading 4, **step 6**).
2. Detection of *KIT* Exon 17 mutation analysis by HRMA showed a false positive rate of 7.9 %; therefore all samples with discordant and variant HRMA calls will be confirmed with Sanger sequencing (Subheading 4, **step 9**).
3. HRMA assay can only be used to determine the presence of mutation in a sample on the basis of variant melt curve as compared to the wild-type control. It is unable to define the exact nature of the mutation causing the variant melt curve.

References

1. Nguyen-Dumont T, Calvez-Kelm FL, Forey N et al (2009) Description and validation of high-throughput simultaneous genotyping and mutation scanning by high-resolution melting curve analysis. Hum Mutat 30:884–890
2. Erali M, Voelkerding KV, Wittwer CT (2008) High resolution melting applications for clinical laboratory medicine. Exp Mol Pathol 85:50–58
3. Vossen RH, Aten E, Roos A et al (2009) High-resolution melting analysis (HRMA): more than

just sequence variant screening. Hum Mutat 30:860–866

4. Montgomery J, Wittwer CT, Palais R et al (2007) Simultaneous mutation scanning and genotyping by high-resolution DNA melting analysis. Nat Protoc 2:59–66
5. Smith BL, Lu CP, Alvarado Bremer JR (2010) High-resolution melting analysis (HRMA): a highly sensitive inexpensive genotyping alternative for population studies. Mol Ecol Resour 10:193–196
6. Wittwer CT (2009) High-resolution DNA melting analysis: advancements and limitations. Hum Mutat 30:857–859
7. Patel KP, Barkoh BA, Chen Z et al (2011) Diagnostic testing for *IDH1* and *IDH2* variants in acute myeloid leukemia an algorithmic approach using high-resolution melting curve analysis. J Mol Diagn 13:678–686
8. Singh RR, Bains A, Patel KP et al (2012) Detection of high-frequency and novel *DNMT3A* mutations in acute myeloid leukemia by high-resolution melting curve analysis. J Mol Diagn 14:336–345
9. Cammenga J, Horn S, Bergholz U et al (2005) Extracellular KIT receptor mutants, commonly found in core binding factor AML, are constitutively active and respond to imatinib mesylate. Blood 106:3958–3961

Chapter 8

Locked Nucleic Acid Probes (LNA) for Enhanced Detection of Low-Level, Clinically Significant Mutations

Khedoudja Nafa, Meera Hameed, and Marie E. Arcila

Abstract

The detection of clinically significant somatic mutations present at low level in a tissue sample represents a challenge in any laboratory. While several high sensitivity methods are described, the incorporation of these new techniques in a clinical lab may be difficult if the technology is not readily available or requires major changes in the workflow of the laboratory. Techniques that are robust and easily adapted to existing laboratory protocols are highly advantageous. In this chapter we describe the use of locked nucleic acid (LNA) probes to modify existing polymerase chain reaction (PCR)-based protocols which can then be sequenced by Sanger sequencing. LNA probes are used to enhance the sensitivity of Sanger sequencing to mutation frequencies below 1 %. The method is robust and is easily incorporated for assessment of any sample with low tumor content or low mutant allele burden.

Key words Locked nucleic acid probes, LNA, LNA clamping, LNA-PCR

1 Introduction

The detection of clinically significant somatic mutations in tumor samples often represents a significant challenge for any laboratory due to excess wild-type DNA from non-neoplastic cells accompanying the tumor. A widely used method for enrichment of tumor cells is manual macrodissection or microdissection. However, adequate enrichment may be precluded by mixed and intimately associated non-neoplastic cells in the tumor or in very small samples not amenable to manual macrodissection. Depending on the volume and type of cases handled by the laboratory, this enrichment modality could pose a major bottleneck in the testing workflow given the labor-intensive nature of the process.

An alternate approach for detection of mutations at low level is the use of higher sensitivity testing methods. Several methods are currently described in the literature which may require major changes in the testing protocols or call for the use of additional instrumentation not necessarily available or easily implemented in

Rajyalakshmi Luthra et al. (eds.), *Clinical Applications of PCR*, Methods in Molecular Biology, vol. 1392, DOI 10.1007/978-1-4939-3360-0_8,

a laboratory. Enrichment methods for low-level mutations that are easily incorporated into existing protocols present a distinct advantage. One of these methods is the locked nucleic acid (LNA) PCR. LNAs are nucleic acid analogs that contain a 2′-O,4′-C-methylene bridge in the ribose moiety. LNA bases can be incorporated into any DNA oligonucleotide, and each LNA base increases the thermal stability of DNA/probe heteroduplexes by 3–8 °C [1, 2]. When oligonucleotides are designed to be complimentary to wild-type DNA, the resulting LNA/DNA hybrids exhibit high thermal stability and wild-type allele amplification is suppressed. If a mutation is present in the same region, due to high specificity the LNA probe does not bind to the mutant allele and this is preferentially amplified [3, 4].

In this chapter we outline the protocol used in our laboratory for the detection of common somatic driver mutations in three genes with significant clinical implications. This includes detection of the *EGFR* T790M mutation, *BRAF* V600E mutation, and *KRAS* mutations in codons 12 and 13 [5, 6].

For the assays described below, we designed three LNA-containing oligonucleotide (oligo) clamps consisting of ten consecutive LNA monomers complementary to the wild-type sequences of *EGFR* (codons 789–791), *BRAF* (codons 598–601), and KRAS (codons 12–14) (*see* Fig. 1, Table 1). In our experience, optimal design of LNA clamps positions the specific codon of interest (where mutation would be expected) in the center of the sequence, with the remaining LNA bases distributed evenly at both ends. PCR master mixes and cycling conditions remain identical to those in the standard assays, except for the addition of the LNA probe to the PCR mixture. Although here we describe our own protocols, the LNA probes can also be adapted to other protocols and techniques. In our clinical assays we run all samples in duplicate with standard and LNA-PCR and each duplicate is sequenced in both forward and reverse directions.

In addition to modification of PCR methods, LNA-enhanced oligonucleotides can be used in a wide range of other applications [4, 7–24]. LNA oligonucleotides can be designed by the user but optimized design can also be provided by the vendor. General guidelines for the design include avoiding self-complementarity and cross-hybridization to other LNA-containing oligonucleotides, keeping GC content between 30 and 60 %, avoiding stretches of 3 or more Gs or Cs and maintaining the length of consecutive LNA stretches at around 4 LNA bases, except when very short (9–10 nt) oligonucleotides are designed and avoiding positioning of probes over palindrome sequences. For clamp design, sequences are made to prevent extension during the PCR reaction which is usually achieved by the introduction of 3′ modifications such as inversed T (3InvdT) or phosphorylation.

A. BRAF V600

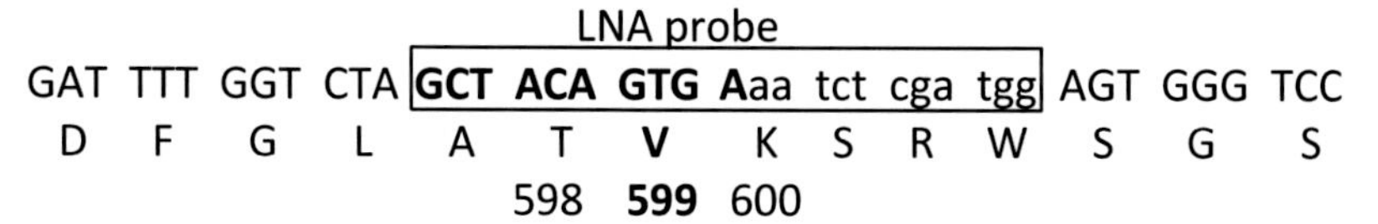

B. EGFR T790

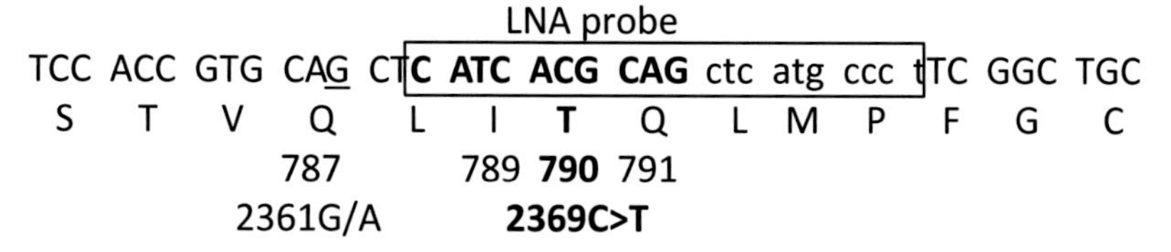

C. KRAS G12-G13

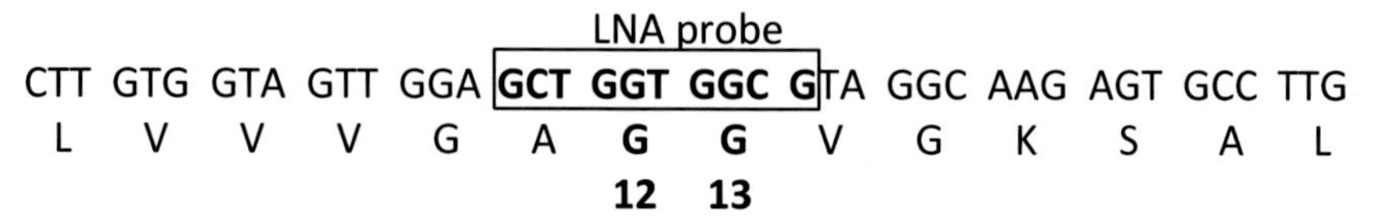

Fig. 1 Location of the LNA probes on genomic sequences of BRAF, EGFR, and KRAS. EGFR T790, BRAF V600, and KRAS G12-G13 are in bold. A common germline EGFR SNP 2361G/A (Q787Q) is indicated. The LNA probes are boxed, the locked nucleotides are bolded, and the unlocked nucleotides at the 3′end (lower case) are added to stabilize the EGFR and BRAF LNA probes to predicted Tm of approximately 80°

Table 1
List of primers and LNA probes

Gene/exon F/R	Primer (5′->3′)	Size (bp)
M13F2[a]	*gtaaaacgacggccagt*	–
M13R2[a]	*caggaaacagctatgacc*	–
EGFR-Ex20-F-M13F2[b] EGFR-Ex20-R-M13R2[b]	gtaaaacgacggccagtc ATTCATGCGTCTTCACCTG caggaaacagctatgaccc ATATCCCCATGGCAAACTC	412
EGFR-LNA-T790M-F[c] probe	+C+A+T+C+A+C+G+C+A+Gctcatgccct/3InvdT/	–
BRAF-15 F-M13F2[b] BRAF-15R-M13R2[b]	gtaaaacgacggccagt TCATAATGCTTGCTCTGATAGG caggaaacagctatgacc GGCCAAAAATTTAATCAGTGG	259
BRAF-LNA-F[c] probe	+G+C+T+A+C+A+G+T+G+Aaatctcgatgg/3InvdT/	
KRAS-F-M13F2[b] KRAS-R-M13F2[b]	gtaaaacgacggccagt GTGTGACATGTTCTAATATAGTCA caggaaacagctatgacc CTGTATCAAAGAATGGTCCTGCAC	259
KRAS LNA-F[c] probe	G+C+T+G+G+T+G+G+C+G/3InvdT/	

[a]M13 primers for sequencing

[b]M13 nucleotide sequence (italics) added to EGFR specific primer for Sanger sequencing

[c]In the LNA oligonucleotide probe, the locked nucleotides are in capital letters followed by the plus (+) sign and the unlocked nucleotides at the 3′end (lower case) are added to stabilize the EGFR and BRAF probes to predicted Tm of 82 and 78 °C, respectively. The addition at the 3′ end of "3InvdT" inhibits the extension of the LNA probe

2 Materials

2.1 Sample Requirements

Any source of DNA is suitable for testing: Archival formalin-fixed paraffin-embedded (FFPE) and fresh tissues, aspirates, or body fluids in cytolite. Tumor content may be as low as 1–2 %. As with other PCR methods, most decalcified material is unsuitable for testing.

2.2 Reagents

Positive (cell lines from ATCC are listed in Table 2) and negative controls DNA.

HotStar Taq DNA Polymerase kit containing 10× PCR buffer and 15 mM $MgCl_2$ (Qiagen).

25 mM $MgCl_2$.

AmpErase® Uracil *N*-glycosylase (UNG) (Applied Biosystem).

GeneAmp® dNTP Blend, 12.5 mM with dUTP (Applied Biosystem).

PCR primers (IDT custom oligonucleotide): working solution at 4 pmol/μl (*see* Tables 3 and 4).

LNA oligonucleotide probes (custom probes from Exiqon).

ExoSAP-IT PCR product cleanup, 100 reactions, 200 μl (Affymetrix, Inc).

DNA sequencing reagents (Applied Biosystem).

Table 2
Positives control cell lines

ATCC Cat#	Name	Gene/mutation
CRL-5908	NCI-H1975	EGFR T790M (exon 20) and L858R (exon 21)
HTB-73	SK-MEL-31	BRAF V600E (exon 15)
CRL-5807	NCI-H358	KRAS 34G>T, G12C (exon 2)

Table 3
Dilution of primers for PCR and sequencing

Primers/oligo (100 μM)	F+R mix (4 pmol/μl)	F+R+LNA mix (4 pmol/μl)	F (4 pmol/μl)	R (4 pmol/μl)
Forward (F) PCR primer	40 μl	40 μl	–	–
Reverse (R)PCR primer	40 μl	40 μl	–	–
LNA oligonucleotide	–	40 μl	–	–
M13F2	–	–	40 μl	–
M13R2	–	–	–	40 μl
Water (up to 1000 μl)	920 μl	880 μl	960 μl	960 μl

Table 4
Composition of PCR reaction mix

Reagent	Standard PCR		LNA-PCR	
	1 Reaction (μl)	6 Reactions (μl)	1 Reaction (μl)	6 Reactions (μl)
PCR water up to 40 μl	**24.25**	145.5	**24.25**	145.5
10× Buffer containing 15 mM $MgCl_2$	**5**	30	**5**	30
dNTP 12.5 mM with dUTP	**4**	24	**4**	24
$MgCl_2$ (25 mM)	**1**	6	**1**	6
PCR primers mix (4 pmol)	**5**	30	**0**	0
PCR primers + LNA probe mix (4 pmol)	**0**	0	**5**	30
AmpErase® Uracil *N*-glycosylase (UNG)	**0.5**	3	**0.5**	3
HotStarTaq Plus Polymerase	**0.25**	1.5	**0.25**	1.5
Total volume	**40**	**240**	**40**	**240**

Thermal cycler.
Agarose gel electrophoresis apparatus.
DNA sequencing reagents.
Automated DNA sequencing equipment.

3 Methods

3.1 PCR Amplification

Each sample is run with both standard protocol and LNA protocol in duplicate.

1. Quick spin the vial containing liquid primer to bring contents to the bottom of the tube.
2. Check Certificate of analysis to find the concentration in μM in the vial (100 μM).
3. To make a working primer solution from the stock of the forward and reverse primers, dilute with water to a final concentration of 4 pmol according to Table 3. Diluted primers are stored at −20 °C for up to 1 year.
4. Make up 40 μl of PCR reaction master mix for each sample according to Table 5. A minimum of 6 reactions for each PCR type are required as follows: 2 for patient sample, 1 positive control, 1 negative control, 1 no-template control (NTC), and 1 extra for pipetting error.
5. Transfer 40 μl of the Amplification Mix to the bottom of each well of the plate

Table 5
DNA input for PCR

Sample concentration	DNA volume (μl)	Water (up to 10 μl)	Total DNA added (ng)
1–25 ng/μl	10	None	10–250
25–100 ng/μl	5	5	125–500
101–200 ng/μl	2	8	202–400
>200 ng/μl	1	9	>200

Table 6
PCR cycling conditions

Temperature	Time	Cycling
Room temperature[a]	10 min	1 Cycle
95 °C[b]	15 min	1 Cycle
94 °C 60 °C 72 °C	45 s 45 s 1 min	42 Cycles
72 °C 4 °C	10 min Hold	1 Cycle

[a]Incubation can be done at room temperature for 10 min or at 50 °C for 2 min (UNG degrades any contaminating PCR products)
[b]Inactivates *N*-glycosylase and activates Taq DNA polymerase

6. Pipet Nuclease-Free Water according to Table 4 into the wells containing the Amplification Mix.
7. For the blank amplification control (no DNA control), pipet 10 μl Nuclease-Free Water (instead of template DNA) into the well containing the Amplification Mix. Mix by pipetting several times.
8. Pipet the template DNA of each sample according to Table 5 into the bottom of the respective well containing Amplification Mix. Mix by pipetting several times.
9. Pipet the control DNA (100–200 ng) into the bottom of the respective well containing Amplification Mix. Mix by pipetting several times.
10. Cap the wells securely and vortex.
11. Spin the plate for 1 min at 657 g.
12. Perform PCR amplification according to Table 6.
13. Check the PCR product intensity by running 5 μl of each PCR reaction with 2 μl of loading dye on a 2 % agarose gel stained with ethidium bromide.

3.2 DNA Sequencing

All PCR products (standard and LNA-PCRs) are sequenced on an ABI 3730 DNA using Big Dye Terminator Chemistry (Applied Biosystems) according to the manufacturer's recommendations. Forward M13F2 and the reverse M13R2 primers are used for sequencing reaction. The limit of detection of the high sensitivity LNA-PCR/sequencing assay is summarized in **Note 1**.

Good forward and reverse sequence data with well-resolved signal peaks should allow reads of almost all of the sequence from both strands. Any nucleotide change in one direction (on one strand) should be confirmed to be present in the other direction (the other strand).

3.3 Data Analysis

3.3.1 Negative Results

A negative mutation call requires the assessment of Sanger sequencing results after standard PCR and LNA-PCR concurrently. As the LNA-PCR suppresses the amplification of wild type, in negative cases where all wild-type template is suppressed, the Sanger sequencing will show a failure pattern that is completely indistinguishable from PCR or sequencing failure for others reason. Concurrent Sanger sequencing results after standard PCR showing a readable sequence with no mutations will serve as the control for failures due to other reasons.

For cases in which the wild-type template is not entirely suppressed, a negative result is indicated by the absence of a nucleotide change in the coding region of interest compared to the electrophoretogram of the reference normal sample for both standard and LNA-PCR. For cases in which the wild-type template is entirely suppressed, a negative result is indicated by an absence of nucleotide changes in the coding region of interest with standard PCR and a failure pattern with the LNA-PCR (Fig. 2).

3.3.2 Positive Results

EGFR T790M, BRAF V600, and KRAS G12-G13 mutations are listed in Table 7.

In samples with high mutant allele content, the mutant peak is readily seen in the standard PCR sequencing (Fig. 3). At very low mutant content, however, a mutant peak may be at the level of the background noise or not detectable. With the addition of an LNA clamp, the vast majority of cases will only show the mutant allele as the wild type is not amplified and therefore not detectable (Fig. 4). Cases with excess amounts of non-mutant cells will show both alleles due to the high proportion of normal DNA template not entirely suppressed by the LNA oligonucleotide during the PCR reaction. In these cases, the wild-type peak is generally much smaller than the mutant. For our assays we require that the mutant peak is at least three times higher than the wild type for an unequivocal positive call (Fig. 5 and **Note 1**).

3.3.3 Equivocal Results and Artifactual Peaks

The high sensitivity LNA-PCR/sequencing assay allows for the detection of very low concentrations of mutant cells down to 0.5 %. Samples with tumor content near this level or below may

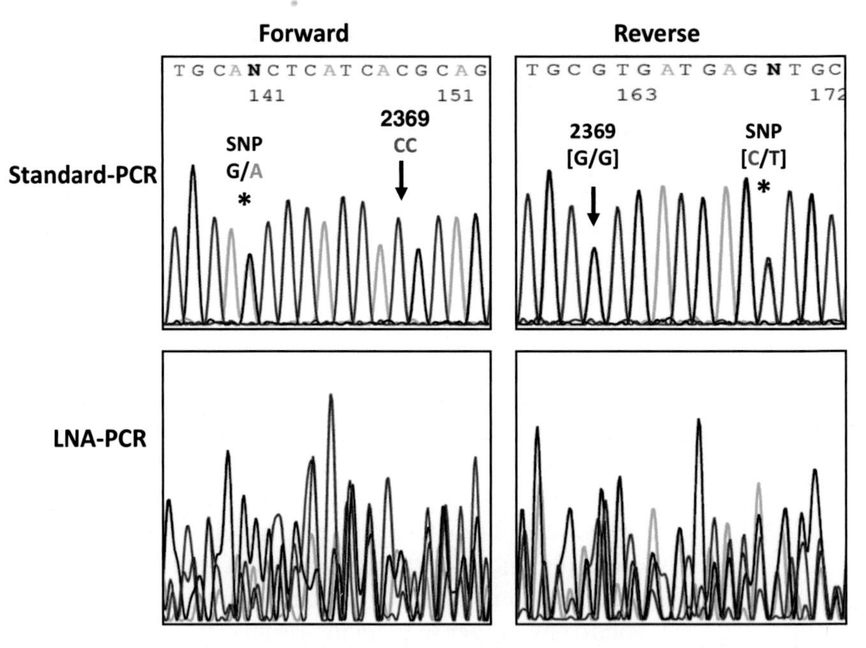

Fig. 2 Electropherogram of an EGFR LNA-PCR sequencing failure. When the wild-type template is entirely suppressed, a negative result is indicated by an absence of change at nucleotide position 2369 with standard PCR and a failure pattern with the LNA-PCR

Table 7
List of mutations of EGFR T790, BRAF V600, and KRAS G12-G13

Gene nucleotide change	Amino acid change
*BRAF 1799*GTG > GCG	Val600Ala, p.V600A
BRAF 1799GTG > GAG	Val600Glu, p.V600E
BRAF 1799GTG > GGG	Val600Gly, p.V600G
EGFR 2369A CG > ATG	Thr858Met, p.T790M
KRAS 34**G**GT > **A**GT	Gly12Ser, p.G12S
KRAS 34**G**GT > **C**GT	Gly12Arg, p.G12R
KRAS 34**G**GT > **T**GT	Gly12Cys, p.G12C
KRAS 35G**G**T > G**A**T	Gly12Asp, p.G12D
KRAS 35G**G**T > G**C**T	Gly12Ala, p.G12A
KRAS 35G**G**T > G**T**T	Gly12Val, p.G12V
KRAS 37**G**GC > **A**GC	Gly13Ser, p.G13S
KRAS 37**G**GC > **C**GC	Gly13Arg, p.G13R
KRAS 37**G**GC > **T**GC	Gly13Cys, p.G13C
KRAS 38G**G**C > G**A**C	Gly13Asp, p.G13D
KRAS 38G**G**C > G**C**C	Gly13Ala, p.G13A
KRAS 38G**G**C > G**T**C	Gly13Val, p.G13V

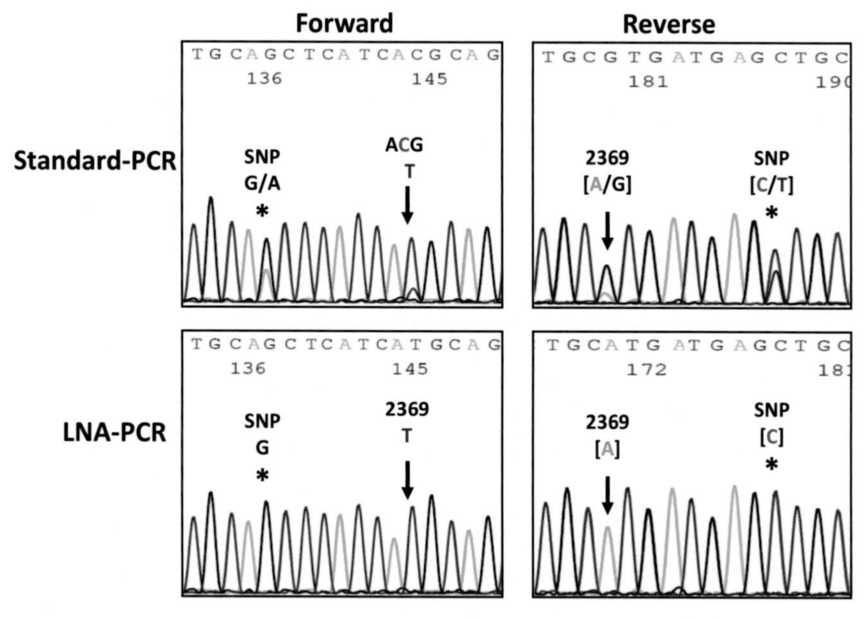

Fig. 3 Electropherogram of an EGFR T790M positive by both standard and LNA-PCR Sequencing. In this sample with high mutant allele content, the wild-type template is entirely suppressed, a positive result is indicated by the presence of a change at nucleotide position 2369 with standard PCR, and only the mutant peak is seen with LNA-PCR

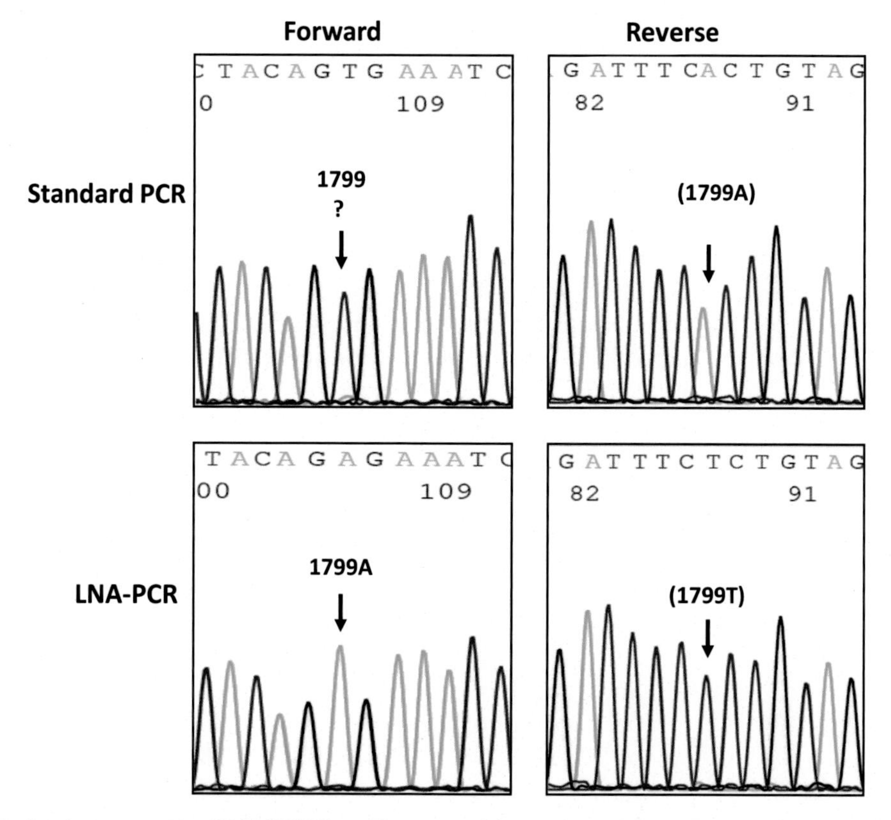

Fig. 4 Electropherogram of a BRAF V600E positive case with very low mutant allele content. In the standard PCR, a mutant peak at the level of the background is suspected. With the LNA-PCR only a mutant peak is seen

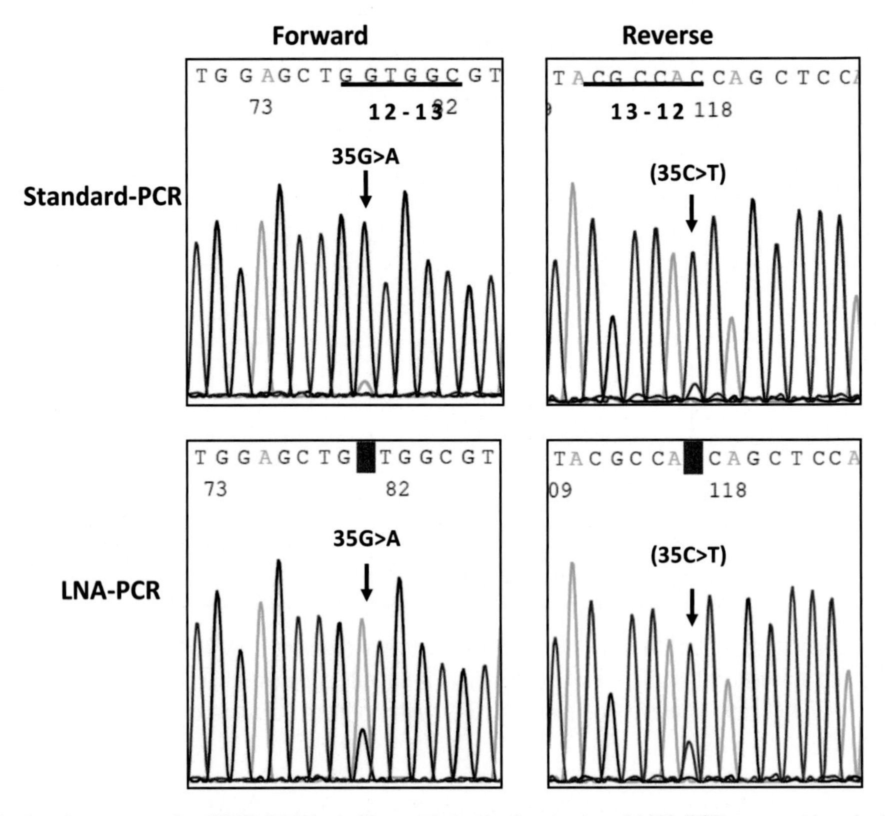

Fig. 5 Electropherogram of a KRAS G12D positive with both standard and LNA-PCR sequencing. In the LNA-PCR the mutant peak is three times higher than the wild type

sometimes show mutant peaks that are not significantly higher that the wild type and which are not fully reproducible in all reactions. Retesting of these samples may resolve some of these difficult calls. In our experience, mutant peaks of the same size as the wild type are often accompanied by several artifactual peaks in the region of the LNA probe and represent pseudo-mutation peaks that are not reproducible in repeat reactions (Fig. 6). As a rule, we have established our cutoff for an unequivocal positive call as mutant to wild-type ratio of 3:1. Please see notes below for further information on artifactual patterns.

4 Notes

1. For the validation of the analytical sensitivity of the standard and the LNA-PCR sequencing, we recommend using a cell line as well as a mutant clinical sample DNA to establish the performance of the assay. For our validation we used dilutions of 50, 25, 12.5, 6, 3, 1.5, 0.75, 0.38, and 0.19 % of mutant DNA for LNA-PCR and dilutions of 50, 25, 12.5, 6, and 3 % for standard PCR sequencing. The addition of the forward

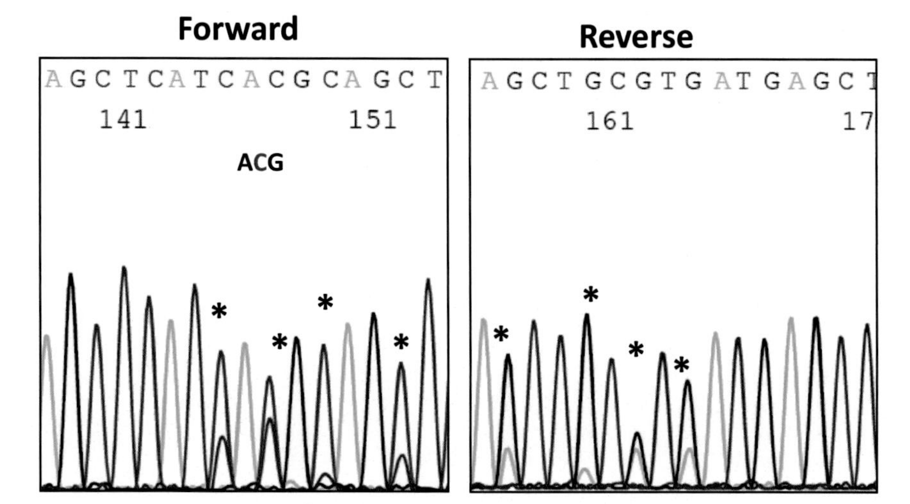

Fig. 6 Electropherogram of an EGFR T790 LNA-PCR sequencing with artifactual mutant peaks in the region bound by the LNA probe (*). On repeat runs, these artifactual peaks are not reproducible

LNA probes improves the sensitivity of EGFR T790M, BRAF, and KRAS from 12.5 % for EGFR T790M and 25 % for KRAS and BRAF V600E to 0.4, 0.2, and 0.1 % in both directions, respectively. At concentrations of mutant DNA of 0.75 % for EGFR T790M, 3 % for BRAF and KRAS as described above, only the mutant peak will be present [5, 6]. For clinical purposes, the technical sensitivity of the assay for all three mutations is conservatively estimated at 1 %.

It should be noted that at extremely low concentrations of mutant allele (generally below 0.1 % for EGFR, 0.19 % for KRAS, and 0.75 % for BRAF), additional artifact peaks (representing low-level errors due to mis-incorporation of Taq polymerase or formalin fixation) are often unmasked, making the assay difficult to interpret. Such a pattern becomes readily recognizable and distinct from true positive results. Repeat runs in these cases do not reproduce the same pattern of mutant peaks, confirming them as artifactual (Fig. 4). Occasionally, sequencing tracings may also be accompanied by a high background throughout. Cases showing a mutant peak in association with this background sometimes yield negative results in repeat runs. These cases should therefore be managed conservatively by repeating the testing to confirm the mutation status.

References

1. Vester B, Lundberg LB, Sorensen MD, Babu BR, Douthwaite S, Wengel J (2004) Improved RNA cleavage by LNAzyme derivatives of DNAzymes. Biochem Soc Trans 32:37–40
2. Pfundheller HM, Sorensen AM, Lomholt C, Johansen AM, Koch T, Wengel J (2005) Locked nucleic acid synthesis. Methods Mol Biol 288:127–146

3. Petersen M, Wengel J (2003) LNA: a versatile tool for therapeutics and genomics. Trends Biotechnol 21:74–81
4. Mouritzen P, Nielsen AT, Pfundheller HM, Choleva Y, Kongsbak L, Moller S (2003) Single nucleotide polymorphism genotyping using locked nucleic acid (LNA). Expert Rev Mol Diagn 3:27–38
5. Arcila M, Lau C, Nafa K, Ladanyi M (2011) Detection of KRAS and BRAF mutations in colorectal carcinoma roles for high-sensitivity locked nucleic acid-PCR sequencing and broad-spectrum mass spectrometry genotyping. J Mol Diagn 13:64–73
6. Arcila ME, Oxnard GR, Nafa K, Riely GJ, Solomon SB, Zakowski MF, Kris MG, Pao W, Miller VA, Ladanyi M (2011) Rebiopsy of lung cancer patients with acquired resistance to EGFR inhibitors and enhanced detection of the T790M mutation using a locked nucleic acid-based assay. Clin Cancer Res 17:1169–1180
7. Orum H, Jakobsen MH, Koch T, Vuust J, Borre MB (1999) Detection of the factor V Leiden mutation by direct allele-specific hybridization of PCR amplicons to photoimmobilized locked nucleic acids. Clin Chem 45:1898–1905
8. Jacobsen N, Fenger M, Bentzen J, Rasmussen SL, Jakobsen MH, Fenstholt J, Skouv J (2002) Genotyping of the apolipoprotein B R3500Q mutation using immobilized locked nucleic acid capture probes. Clin Chem 48:657–660
9. Jacobsen N, Bentzen J, Meldgaard M, Jakobsen MH, Fenger M, Kauppinen S, Skouv J (2002) LNA-enhanced detection of single nucleotide polymorphisms in the apolipoprotein E. Nucleic Acids Res 30:e100
10. Simeonov A, Nikiforov T (2002) Single nucleotide polymorphism genotyping using short, fluorescently labeled locked nucleic acid (LNA) probes and fluorescence polarization detection. Nucleic Acids Res 30:e91
11. Michikawa Y, Suga T, Ohtsuka Y, Matsumoto I, Ishikawa A, Ishikawa K, Iwakawa M, Imai T (2008) Visible genotype sensor array. Sensors 8:2722–2735
12. Choleva Y, Norholm M, Pedersen S, Mouritzen P, Hoiby P, Nielsen AT, Moller S, Jacobsen M, Kongsbak L (2001) Multiplex SNP genotyping using Locked Nucleic Acid and microfluidics. JALA 6:92–97
13. Latorra D, Campbell K, Wolter A, Hurley JM (2003) Enhanced allele-specific PCR discrimination in SNP genotyping using 3′ locked nucleic acid (LNA) primers. Hum Mutat 22:79–85
14. Di Giusto DA, King GC (2004) Strong positional preference in the interaction of LNA oligonucleotides with DNA polymerase and proofreading exonuclease activities: implications for genotyping assays. Nucleic Acids Res 32:e32
15. Rupp J, Solbach W, Gieffers J (2006) Single-nucleotide-polymorphism-specific PCR for quantification and discrimination of Chlamydia pneumoniae genotypes by use of a "locked" nucleic acid. Appl Environ Microbiol 72:3785–3787
16. Johnson MP, Haupt LM, Griffiths LR (2004) Locked nucleic acid (LNA) single nucleotide polymorphism (SNP) genotype analysis and validation using real-time PCR. Nucleic Acids Res 32:e55
17. Ugozzoli LA, Latorra D, Puckett R, Arar K, Hamby K (2004) Real-time genotyping with oligonucleotide probes containing locked nucleic acids. Anal Biochem 324:143–152
18. Grannemann S, Landt O, Breuer S, Blomeke B (2005) LightTyper assay with locked-nucleic-acid-modified oligomers for genotyping of the toll-like receptor 4 polymorphisms A896G and C1196T. Clin Chem 51:1523–1525
19. Hu Y, Le Leu RK, Young GP (2009) Detection of K-ras mutations in azoxymethane-induced aberrant crypt foci in mice using LNA-mediated real-time PCR clamping and mutant-specific probes. Mutat Res 677:27–32
20. Senescau A, Berry A, Benoit-Vical F, Landt O, Fabre R, Lelievre J, Cassaing S, Magnaval JF (2005) Use of a locked-nucleic-acid oligomer in the clamped-probe assay for detection of a minority Pfcrt K76T mutant population of Plasmodium falciparum. J Clin Microbiol 43:3304–3308
21. Ren XD, Lin SY, Wang X, Zhou T, Block TM, Su YH (2009) Rapid and sensitive detection of hepatitis B virus 1762 T/1764A double mutation from hepatocellular carcinomas using LNA-mediated PCR clamping and hybridization probes. J Virol Methods 158:24–29
22. You Y, Moreira BG, Behlke MA, Owczarzy R (2006) Design of LNA probes that improve mismatch discrimination. Nucleic Acids Res 34:e60
23. McTigue PM, Peterson RJ, Kahn JD (2004) Sequence-dependent thermodynamic parameters for locked nucleic acid (LNA)-DNA duplex formation. Biochemistry 43:5388–5405
24. Levin JD, Fiala D, Samala MF, Kahn JD, Peterson RJ (2006) Position-dependent effects of locked nucleic acid (LNA) on DNA sequencing and PCR primers. Nucleic Acids Res 34:e142

Chapter 9

Genotyping of Frequent Mutations in Solid Tumors by PCR-Based Single-Base Extension and MassARRAY Analysis

Alberto Paniz Mondolfi and Rajesh R. Singh

Abstract

Over the last decade, cancer genome sequencing has revealed in detail the genomic landscapes of an increasing number of common solid human tumors. This has greatly impacted the clinical care of cancer patients based on the fact that many of these tumors exhibit activating mutations in driver genes that are prone to target therapy, largely impacting cancer management strategies. Genomic heterogeneity of tumors is becoming an increasingly recognized phenomenon relevant to genome-based medicine. Because a large number tumors may display several mutations at the same time, multiplexing has become the preferred approach to reveal the mutational landscape in patient samples and to better design a targeted approach to their illness.

Key words PCR, MassARRAY, SNP, Solid tumor, Genotyping

1 Introduction

Over the past decade the advent of numerous targeted therapies has risen as a consequence of an in-depth knowledge on genomic human cancer landscapes, thus revolutionizing cancer patient management. Personalized cancer therapy, defined as individualized treatment based on the molecular profile of an individual tumor, yielding precise and predictive response is already having a critical impact on clinical care of cancer patients [1, 2]. Genetic heterogeneity on the other hand is emerging as a relevant concept in tumorigenesis because of its importance not only in tumor biology and evolution but also in its approach to personalized treatment. Intratumoral heterogeneity has been recognized for years. At the genomic level, tumors may present with a wide variety of aberrations, including mutations in coding sequences, gene amplifications, and translocations. In addition, intermetastatic, intrametastatic, and interpatient heterogeneity has been recognized as well [1]. Genomic heterogeneity plays an important role in how a tumor responds to a specific agent consequently

Rajyalakshmi Luthra et al. (eds.), *Clinical Applications of PCR*, Methods in Molecular Biology, vol. 1392,
DOI 10.1007/978-1-4939-3360-0_9,

impacting patient care [1]. The detection of specific aberrations linked to specific cancer types may serve as molecular markers for diagnosis, prognosis, prediction, and disease monitoring as well as targets for development of therapeutic agents [1, 2]. The recognition of activating mutations in driver genes encoding for example protein kinases has impacted genome-based medicine [1, 3, 4]. A number of therapeutic agents have been developed for several genes involved in the pathogenesis of various solid tumor malignancies, including lung adenocarcinomas, gastrointestinal stromal tumors (GISTs), colorectal carcinomas, melanoma, and endometrial carcinomas [1, 3, 4].

The vast majority of solid tumor testing is performed on formalin-fixed paraffin-embedded tissue (FFPET), as well as fine needle aspirates (smears, cell blocks). However, a common challenge is that DNA quantity is typically limited and the testing of multiple targets per patient sample (i.e., *RAS, BRAF, and PIK3CA*), which is a common practice, is not always possible. Although Sanger sequencing remains the Gold standard for mutation detection, the technique is labor intensive and is unable to interrogate more than one target at a time, in addition to its relatively high cost. The low frequency of some mutations results in great amount of resources spent on sequencing wild-type patient samples; thus our institution developed a nine-well multiPLEX Pro DNA-based primer extension assay to overcome such issue by using the MassARRAY [5, 6] platform to assess the mutation status of 85 hotspot regions of 13 genes: *AKT1, AKT2, AKT3, BRAF, GNAS, GNAQ, IDH1, IDH2, KRAS, MET, NRAS, PIK3CA, and RET.* The procedure for this assay will be described here.

The use of matrix-assisted laser desorption/ionization (MALDI) mass spectrometry (MS) in the analysis of nucleic acids has been implemented for decades [5, 6]. Even though with some limitations like for example the less stable nature of nucleic acids ions in the vacuum and the limited nucleotide length of resolution, its use in DNA sequence analysis has been groundbreaking. Because the molecular mass limit for UV-laser excitation of DNA is approximately 15 kDa, increasing ion fragmentation above this molecular mass may decrease detection sensitivity [7–9]. This is the reason why MS is more suitable for resequencing and interrogating known sequences such as SNPs rather than for de novo sequencing [10, 11]. Multiplex genotyping using MALDI-TOF MS coupled to single-base primer extension (SBE) using standard ddNTP terminators has been a valid approach; however the small separation of SBE yields has been a limiting factor that has been reduced by incorporating mass-modified terminators through the iPLEX assay [11].

The MassARRAY system is based on multiPLEX Pro PCR followed by a single-base primer extension reaction [6]. The primer extension products are analyzed using Matrix-assisted laser desorption/ionization time-of-flight mass spectrometry or MALDI-TOF

MS. Once the chip is placed into mass spectrometer, each spot is subjected to MALDI-TOF MS. Here a matrix is used to protect the biomolecule (DNA) from being destroyed by direct laser beam under vacuum and to facilitate vaporization and ionization [5]. The biomolecule is ionized via laser pulses. Molecules are separated according to mass-to-charge ratio (*m/z*). The ionized molecules "fly" through the vacuum tube toward the detector. Detection of the ion is based on its flight time [5] to the end of the tube, with low mass molecules reaching the detector first. The time-of-flight is converted to a spectrogram that plots mass versus intensity. Mass-modified terminator nucleotides incorporated during the single-base extension are used to distinguish different genotypes [5, 6].

2 Materials

2.1 Reagents, Special Supplies, and Equipments

1. HPLC Grade Water.
2. Non-Skirted 0.2 ml 96-well reaction plates.
3. Sequenom adhesive PCR plate sealing sheets.
4. Corning 96 well microplate aluminum tape.
5. Roller for Microseal film.
6. 0.2 ml PCR reaction tubes with caps (8 Strips).
7. PCR Optical Caps (8 strips).
8. Autoclaved 1.5 ml PCR reaction tubes (flat top).
9. Biohazard bag with holder.
10. 12-Channel Pipette.
11. Rainin Green-Pack Tip Rack Refill.
12. Adjustable pipettors, 1–1000 (Pre-PCR only).
13. Adjustable pipettors, 0.5–1000 (Post-PCR only).
14. Non-filtered pipette tips (0.5–200).
15. Sterile Aerosol-filtered pipette tips (1–1000).
16. Thermal cycler PE 9700 or Hybaid or ABI2720.
17. iPLEX Pro Genotyping Kit.
18. 10× PCR Buffer.
19. 25 mM $MgCl_2$.
20. 25 mM dNTP Mix.
21. PCR Enzyme (5 U/μl).
22. SAP Buffer.
23. SAP Enzyme (1.7 U/μl).
24. iPLEX Pro Buffer.
25. iPLEX Pro Termination Mix.

26. iPLEX Pro Enzyme.
27. Three-Point Calibrant.
28. SpectroCHIP.
29. Clean Resin.
30. 96-Well dimple plate.
31. Extension Primer Panel (linear regression mixture).
32. 1 μM PCR Primer Mix [9].
33. 96-Well Plate rotator.
34. Desiccant.
35. Nanodispenser RS1000.
36. Ethanol (50 and 100 %).
37. 0.1 N Sodium Chloride.
38. Aerosol Duster.

2.2 Primers and Probes Utilized

1. Extension primers are ordered at 100 nmol scale while the PCR primers are ordered at 25 nmol scale. Primers are reconstituted to 100 μM (PCR primers) and 400 μM (Extension primers) stock and stored at −20 °C for 6 months. The sequences of primers used for the target region amplification and extension in the nine-well panel are described in Table 1.
2. Adjust PCR primers to working concentration of 1 μM depending on the volume of working mix being made. If total of 500 μl of volume is required for each mixes or each well, add 5 μl of each PCR primers (PCR primers 1 and 2) for that well in a final volume of 500 μl solution. Change working primers every 6 months or before if there are indications of problems.
3. Adjust Extension primers according to linear regression mixture (putting primers according to unextended Primer (UEP) mass lowest to highest and adding 14 μl of the top half and 28 μl of the bottom half with increasing mass). Bring the final volume to 800 μl with HPLC grade water.

2.3 Extension Primer Adjustments

1. Fill in the number of primers and the total volume required for each of the panels or wells. Copy the well, the SNP ID, and the Unextended Primer Mass or UEP Mass from the excel file and sort according to the UEP Mass.
2. Divide each well into low mass and high mass based on the UEP mass and the number of SNPs in each well evenly and add 14 μl of the low mass primers and 28 μl of high mass primers as shown in the worksheet. Bring the volume up to 800 μl with HPLC water.

Table 1
Amplification and extension primers used in the nine-well assay format

Well	SNP_ID	PCR primer 1	PCR primer 2	EXTENSION primer
W1	AKT1_K179M_A536C	ACGTTGGATGTGATCCTGG TGAAGGAGAAG	ACGTTGGATGTCCTTGGCCA CGATGACTTC	CCGCTACTACGCCATGA
W1	AKT3_E17K_G49K	ACGTTGGATGAGCCATCTGT CTTCAAAAGG	ACGTTGGATGTTAACATGCGT GCTTTCCTC	GGCCTCCAGTTTTTATATATT
W1	BRAF_D594_1781A	ACGTTGGATGCCACTCCATC GAGATTTCAC	ACGTTGGATGTCTTCATGAA GACCTCACAG	ACTGTAGCTAGACCAAAA
W1	BRAF_G464_1391G	ACGTTGGATGTGGGCAGAT TACAGTGGGAC	ACGTTGGATGACTTACCATGC CACTTTCCC	CAGTGGGACAAAGAATTG
W1	GNAQ_Q209_2_A	ACGTTGGATGAGAGGTGAC ATTTTCAAAGC	ACGTTGGATGACCTTGCAGAAT GGTCGATG	ATTTTCTTCTCTCTGACCTT
W1	GNAS_Q227_C651	ACGTTGGATGGTGTCCTGA CTTCTGGAATC	ACGTTGGATGGGTAACAGTTG GCTTACT	ACCAAGTTCCAGGTGGACA AAGT
W1	IDH1-R132_3_T	ACGTTGGATGAATATCCCCC GGCTTGTGAG	ACGTTGGATGACATGACTTACT TGATCCCC	AAACCTATCATCATAGGTCG
W1	KRAS_Q61_183A	ACGTTGGATGCATGTACTG GTCCCTCATTG	ACGTTGGATGTGGAGAAACCTG TCTCTTGG	GGTCCCTCATTGCACTGTAC TCCTC
W1	KRAS13-2_G	ACGTTGGATGAGGCCTGCT GAAAATGACTG	ACGTTGGATGCTGTATCGTCAA GGCACTCT	CTTGTGGTAGTTGGAGCTG GTG
W1	MET_N375S	ACGTTGGATGCCATGTGTG CATTCCCTATC	ACGTTGGATGGCTGGAGACATC TCACATTG	CAAATATGTCAACGACTTCT TCA
W1	MET_T1010I_C3029T	ACGTTGGATGATAGGCTTGT AAGTGCCCGA	ACGTTGGATGAGTAGCTCGGTA GTCTACAG	TGTAAGTGCCCGAAGTGTAA GCCCAA
W1	PIK3CA_E545_1635G	ACGTTGGATGTACACGAGAT CCTCTCTCTG	ACGTTGGATGTAGCACTTACC TGTGACTCC	CGAGATCCTCTCTCTGAAAT CACTGA

(continued)

Table 1
(continued)

Well	SNP_ID	PCR primer 1	PCR primer 2	EXTENSION primer
W1	PIK3CA_F909L	ACGTTGGATGGCAGCCATTGACCTGTTTAC	ACGTTGGATGTTGTGACGATCTCCAATTCC	GTGCTGGATACTGTGTAGCTACCTT
W2	AKT3_G171R_G511A	ACGTTGGATGTAGGTAAAGGCACTTTTGGG	ACGTTGGATGTCTTCAGAATCTTCATAGC	TTTTGGTTCGAGAGAAGGCAAGT
W2	GNAS_R201_C601	ACGTTGGATGCTTTGGTGAGATCCATTGAC	ACGTTGGATGGGTCTCAAAGATTCCAGAAG	TCAGGACCTGCTTCGCTGC
W2	KRAS_Q61_181C	ACGTTGGATGCATGTACTGGTCCCTCATTG	ACGTTGGATGTGGAGAAACCTGTCTCTTGG	TCATTGCACTGTACTCCTCTT
W2	MET_H1112Y_C3334T	ACGTTGGATGACGCAGTGCTAACCAAGTTC	ACGTTGGATGGCCATCATTGTCCAACAAAG	GGGCATTTTGGTTGTGTATAT
W2	MET_M1268T_T3803C	ACGTTGGATGAAAACAGGTGCAAAGCTGCC	ACGTTGGATGCATCTGACTTGGTGGTAAAC	AGCTGCCAGTGAAGTGGA
W2	MET_N848S	ACGTTGGATGTTCAACCGTCCTTGGAAAAG	ACGTTGGATGGATATTGAGACAACACCAGC	AAGTAATAGTTCAACCAGATCAGA
W2	MET_R988C	ACGTTGGATGGCCTATCCAAATGAGGAGTG	ACGTTGGATGCTCTGTTTTAAGATCTGGGC	TACTCTTGCATCGTAGC
W2	MET_Y1253D_T3757G	ACGTTGGATGGCAGCTTTGCACCTGTTTTG	ACGTTGGATGGTTGCTGATTTTGGTCTTGC	CTGTTTTGTTGTGTACACTAT
W2	PIK3CA_E418K	ACGTTGGATGATTTCCCCATGCCAATGGAC	ACGTTGGATGTGGGGAAGAAAAGTGTTTTG	CCATGCCAATGGACAGTGTT
W2	PIK3CA_E453K	ACGTTGGATGGGATTTGATCCAGTAACACC	ACGTTGGATGGGCTTTGAATCTTTGGCCAG	CCAATAGGGTTCAGCAAATCTT
W2	PIK3CA_E542_1625A	ACGTTGGATGTAGCACTTACCTGTGACTCC	ACGTTGGATGGCAATTTCTACACGAGATCC	AAATCTTTCTCCTGCTCAGTGATT

W2	PIK3CA_K111N	ACGTTGGATGAAGTAATTG AACCAGTAGGC	ACGTTGGATGGTATCATACC AATTTCTCG	GGCAACCGTGAAGAAAA
W2	PIK3CA_M1043V	ACGTTGGATGTCCATTTTT GTTGTCCAGCC	ACGTTGGATGAACTGAGCAAG AGGCTTTGG	CATGATGTGCATCATTCA
W2	PIK3CA_N345K	ACGTTGGATGAAAAATTCTT TGTGCAACC	ACGTTGGATGGCATCAGCATTT GACTTTACC	AAAATTCTTTGTGCAACCTAC GTGAA
W3	AKT1_E17K_G49A	ACGTTGGATGTTGAGGAGG AAGTAGCGTGG	ACGTTGGATGTAGAGTGTGCG TGGCTCTCA	CCAGGTCTTGATGTACT
W3	AKT1_G173R_G517C	ACGTTGGATGCACTTTCGGC AAGGTGATCC	ACGTTGGATGTGAGGATCTTCA TGGCGTAG	GGTGAAGGAGAAGGCCACA
W3	BRAF_G466_1397G	ACGTTGGATGACTTACCAT GCCACTTTCCC	ACGTTGGATGTGGGCAGATTA CAGTGGGAC	CCTTGTAGACTGTTCCAAA TGAT
W3	BRAF_K601N_A	ACGTTGGATGGGTGATTTT GGTCTAGCTAC	ACGTTGGATGATGGATCCAGA CAACTGTTC	TGATTTTGGTCTAGCTACA GTGAA
W3	MET_H1124D_ C3370G	ACGTTGGATGATCATGGGA CTTTGTTGGAC	ACGTTGGATGATGCCACTTACT GTTCAAGG	GGACAATGATGGCAAGAA AATT
W3	MET_Y1248C_A3743G	ACGTTGGATGGCAGCTTTGC ACCTGTTTTG	ACGTTGGATGGTTGCTGATTT TGGTCTTGC	TGTGTACACTATAGTATTCT TTATCA
W3	NRAS_G13_G38	ACGTTGGATGAGTGGTTCTG GATTAGCTGG	ACGTTGGATGGACTGAGTACA AACTGGTGG	CAGTGCGCTTTTCCCAACA
W3	PIK3CA_E110K	ACGTTGGATGAAGTAATTGA ACCAGTAGGC	ACGTTGGATGGTATCATACCA ATTTCTCG	TAATTGAACCAGTAGGCAAC CGTGAA
W3	PIK3CA_P539R	ACGTTGGATGCTTACCTGT GACTCCATAG	ACGTTGGATGGCTCAAAGCA ATTTCTACAC	TGCTCAGTGATTTCAGAGAGA
W3	PIK3CA_Q060K	ACGTTGGATGAAGAAGCAAG AAAATACCCC	ACGTTGGATGGCTTCTTGAGT AACACTTACG	GAAAATACCCCCTCCAT
W3	PIK3CA_R088Q	ACGTTGGATGGGTTGAAAAA GCCGAAGGTC	ACGTTGGATGTACTCAAGAAGC AGAAAGGG	AGCCGAAGGTCACAAAGT

(continued)

Table 1
(continued)

Well	SNP_ID	PCR primer 1	PCR primer 2	EXTENSION primer
W3	PIK3CA_T1025_3073A	ACGTTGGATGGGAATGCCAGAACTACAATC	ACGTTGGATGGCCTCTTGCTCAGTTTTATC	ACATTGCATACATTCGAAAG
W4	AKT2_G175R_G523C	ACGTTGGATGAGGGAACCTTTGGCAAAGTC	ACGTTGGATGCAGGATCTTCATGGCGTAGT	TGCGGGAGAAGGCCACT
W4	BRAF_G466R_1396_GC	ACGTTGGATGTGGGCAGATTACAGTGGGAC	ACGTTGGATGACTTACCATGCCACTTTCCC	TGGGACAAAGAATTGGATCT
W4	GNAQ_Q209_3_A	ACGTTGGATGAGAGGTGACATTTTCAAAGC	ACGTTGGATGACCTTGCAGAATGGTCGATG	ATTTTCTTCTCTCTGACCT
W4	MET_H1112_3335	ACGTTGGATGGCACAGTGAATTTTCTTGCC	ACGTTGGATGTTTTGCACAGGGCATTTTGG	TGTCCAACAAAGTCCCA
W4	MET_Y1248_T3742	ACGTTGGATGGTTGCTGATTTTGGTCTTGC	ACGTTGGATGGCAGCTTTGCACCTGTTTTG	TGGTCTTGCCAGAGACATG
W4	NRAS_Q61_C181	ACGTTGGATGCCTGTTTGTTGGACATACTG	ACGTTGGATGTCGCCTGTCCTCATGTATTG	TTGGACATACTGGATACAGCTGGA
W4	PIK3CA_A1046V	ACGTTGGATGAACTGAGCAAGAGGCTTTGG	ACGTTGGATGTCCATTTTTGTTGTCCAGCC	TTCATGAAACAAATGAATGATG
W4	PIK3CA_C420R_2	ACGTTGGATGGTTTTATAATTTAGACTAGTG	ACGTTGGATGAGTTTATATTTCCCCATGCC	TCTTTGTTTTTTAAGGAACAC
W4	PIK3CA_Q546_1636C	ACGTTGGATGTAGCACTTACCTGTGACTCC	ACGTTGGATGGCAATTTCTACACGAGATCC	CTCCATAGAAAATCTTTCTCCT
W4	PIK3CA_S405F	ACGTTGGATGCATTCCTGATCTTCCTCGTG	ACGTTGGATGACCTCTTTAGCACCCTTTCG	TGCTGCTCGACTTTGCCTTT
W4	PIK3CA_Y1021C_3062	ACGTTGGATGGGAATGCCAGAACTACAATC	ACGTTGGATGGCCTCTTGCTCAGTTTTATC	TACAATCTTTTGATGACATTGCAT
W5	AKT2_E17K_G49A	ACGTTGGATGTACCGTGGCCTCCAGGTCTT	ACGTTGGATGCACTCAACCTTGTCCTAACC	CCTCCAGGTCTTGATGTATT

W5	BRAF_K601E_AG	ACGTTGGATGATGGATCCAG ACAACTGTTC	ACGTTGGATGATAGGTGATTTT GGTCTAGC	CCCACTCCATCGAGATT
W5	GNAQ_Q209_1_C	ACGTTGGATGGCAGTGTATC CATTTTCTTC	ACGTTGGATGACCTTGCAGAAT GGTCGATG	TTTCTTCTCTCTGACCTTT
W5	IDH1-R132_1_C	ACGTTGGATGACATGACTTA CTTGATCCCC	ACGTTGGATGAATATCCCCCGG CTTGTGAG	CTTACTTGATCCCCATAAG CATGAC
W5	IDH2-R172-1-A	ACGTTGGATGTGGCCTACCT GGTCGCCAT	ACGTTGGATGAAAACATCCCA CGCCTAGTC	GCCTACCTGGTCGCCATGG GCGTGCC
W5	KRAS_Q61_182A	ACGTTGGATGCATGTACTGG TCCCTCATTG	ACGTTGGATGTGGAGAAACCTG TCTCTTGG	CCCTCATTGCACTGTACT CCTCT
W5	NRAS_G12_G34	ACGTTGGATGGACTGAGTAC AAACTGGTGG	ACGTTGGATGAGTGGTTCTG GATTAGCTGG	AACTGGTGGTGGTTGG AGCA
W5	NRAS_Q61_A182	ACGTTGGATGCCTGTTTGTT GGACATACTG	ACGTTGGATGTCGCCTGTCC TCATGTATTG	TGGACATACTGGATACAG CTGGAC
W5	PIK3CA_C420R	ACGTTGGATGGTTTTATAATT TAGACTAGTG	ACGTTGGATGAGTTTATATTTC CCCATGCC	TTCTTTGTTTTTTAAGG AACAC
W5	PIK3CA_M1043I_ G3129	ACGTTGGATGTCCATTTTTG TTGTCCAGCC	ACGTTGGATGAACTGAGCAAG AGGCTTTGG	TTGGAGTATTTCATGAAA CAAAT
W5	PIK3CA_Q546_1637A1	ACGTTGGATGGCAATTTCTA CACGAGATCC	ACGTTGGATGTAGCACTTACCT GTGACTCC	TCTCTGAAATCACTGAGC
W5	RET_M918T	ACGTTGGATGCATCACTTTG CGTGGTGTAG	ACGTTGGATGTCTTTAGGGTCG GATTCCAG	AAAGGGATTCAATTGCC
W6	BRAF469-1	ACGTTGGATGACTTACCATG CCACTTTCCC	ACGTTGGATGTGGGCAGATTAC AGTGGGAC	CCACTTTCCCTTGTAGA CTGTTC
W6	BRAF600-2	ACGTTGGATGTGATGGGAC CCACTCCATCG	ACGTTGGATGTCTTCATGAAGAC CTCACAG	TGGGACCCACTCCATCG AGATTTC
W6	EGFR858-2	ACGTTGGATGACGTACTGGT GAAAACACCG	ACGTTGGATGTCTTTCTCTTCC GCACCCAG	GATCACAGATTTTGGGC

(continued)

Table 1 (continued)

Well	SNP_ID	PCR primer 1	PCR primer 2	EXTENSION primer
W6	KRAS12-2	ACGTTGGATGAGGCCTGCTGAAAATGACTG	ACGTTGGATGCTGTATCGTCAAGGCACTCT	cTGTGGTAGTTGGAGCTG
W6	KRAS146-1	ACGTTGGATGGGCTCAGGACTTAGCAAGAA	ACGTTGGATGTTCAGTGTTACTTACCTGTC	ATGGAATTCCTTTTATTGAAACATCA
W6	NRAS_G13_G37	ACGTTGGATGGACTGAGTACAAACTGGTGG	ACGTTGGATGAGTGGTTCTGGATTAGCTGG	GGTGGTGGTTGGAGCAGGT
W6	NRAS_Q61_A183	ACGTTGGATGTCGCCTGTCCTCATGTATTG	ACGTTGGATGCCTGTTTGTTGGACATACTG	ATGGCACTGTACTCTTC
W6	PIK3CA1047-1	ACGTTGGATGTCCATTTTTGTTGTCCAGCC	ACGTTGGATGAACTGAGCAAGAGGCTTTGG	GTTGTCCAGCCACCATGAT
W6	PIK3CA542-1	ACGTTGGATGGCAATTTCTACACGAGATCC	ACGTTGGATGTAGCACTTACCTGTGACTCC	TTCTACACGAGATCCTCTCTCT
W7	BRAF600-1	ACGTTGGATGATGGGACCCACTCCATCGAG	ACGTTGGATGTCTTCATGAAGACCTCACAG	CACTCCATCGAGATTTCA
W7	IDH1-R132_2_G	ACGTTGGATGAATATCCCCCGGCTTGTGAG	ACGTTGGATGACATGACTTACTTGATCCCC	GGGTAAAACCTATCATCATAGGTC
W7	IDH2-R172-2-G	ACGTTGGATGAAAACATCCCACGCCTAGTC	ACGTTGGATGTGGCCTACCTGGTCGCCAT	ACCAAGCCCATCACCATTGGCA
W7	KRAS13-1	ACGTTGGATGCTGTATCGTCAAGGCACTCT	ACGTTGGATGAGGCCTGCTGAAAATGACTG	gaggGGCACTCTTGCCTACGC
W7	KRAS146-2	ACGTTGGATGTTCAGTGTTACTTACCTGTC	ACGTTGGATGGGCTCAGGACTTAGCAAGAA	ACTTACCTGTCTTGTCTTT
W7	NRAS_G12_G35	ACGTTGGATGAGTGGTTCTGGATTAGCTGG	ACGTTGGATGGACTGAGTACAAACTGGTGG	CGCTTTTCCCAACACCA

W7	PIK3CA_Y1021_3061T	**ACGTTGGATGGGAATGCCAG AACTACAATC**	**ACGTTGGATGGCCTCTTGCTCA GTTTTATC**	**AACTACAATCTTTTGATGA CATTGCA**
W7	PIK3CA1049-1	**ACGTTGGATGCCAATCCATTT TTGTTGTCC**	**ACGTTGGATGTGAGCAAGAGG CTTTGGAGT**	**aatcaATTTTTGTTGTCCA GCCAC**
W7	PIK3CA545-2	**ACGTTGGATGTAGCACTTACCT GTGACTCC**	**ACGTTGGATGGCAATTTCTACA CGAGATCC**	**ATAGAAAATCTTTCTC CTGC**
W8	BRAF_L597R_1790TG	**ACGTTGGATGGACCCACTCCA TCGAGATTT**	**ACGTTGGATGTCTTCATGAAGA CCTCACAG**	**CGAGATTTCACTGTA GCT**
W8	IDH2-R172-3-G	**ACGTTGGATGAGGTCAGTGGA TCCCCTCTC**	**ACGTTGGATGGGACCAAGCCC ATCACCATT**	**CTGGTCGCCATGGG CGTG**
W8	KRAS12-1	**ACGTTGGATGCTGTATCGTCA AGGCACTCT**	**ACGTTGGATGAGGCCTGCTGAA AATGACTG**	**ACTCTTGCCTACGCCAC**
W8	PIK3CA1047-2	**ACGTTGGATGAACTGAGCAAG AGGCTTTGG**	**ACGTTGGATGTCCATTTTTGTT GTCCAGCC**	**cagGAAACAAATGAATGA TGCAC**
W8	PIK3CA545-1	**ACGTTGGATGGCAATTTCTACA CGAGATCC**	**ACGTTGGATGTAGCACTTACCT GTGACTCC**	**TCCTCTCTCTGAAATCACT**
W9	PIK3CA_Q546_G_3	**ACGTTGGATGGCAATTTCTACA CGAGATCC**	**ACGTTGGATGTAGCACTTACCT GTGACTCC**	**TCTGAAATCACTGAGCA**

The sequences of the primers used for amplification of areas of interest in the 11 genes (PCR primers 1 and PCR primers 2) and the extension primers used in the nine wells are listed

3. To perform primer adjustment on the Mass Spectrometer, add 1 μl of each of the mixes in 49 μl of HPLC water in 96-Well non-skirted PCR plate.
4. Follow the protocol for Dispensation and Mass Spectrometry for dispensation of the samples onto the SpectroChip and for mass spectrometry analysis.

2.4 Interpretation Algorithm

Spectra of each assays in a well is compared against the negative control for all patients. Calls made by the SpectroType Software are based on the height of each peak. Manual calls are made based on extra peaks lining up with individual nucleotides. Extra mutant peaks will show decrease in the wild-type peaks.

3 Methods

3.1 PCR Amplification: Addition of DNA to Reaction Plates, Mix to Reaction Plates

1. Clean the hood with 70 % ethanol and with RNASE/DNASE away, wiping down all the areas of the hood along with the pipettes.
2. Remove PCR reagents from the freezer except for the Enzyme and allow the reagents to thaw inside the hood.
3. Resuspend and homogenize the DNA samples by quick vortex and spin.
4. Make DNA dilutions for samples to have a final concentration of 5 ng/μl.
5. In order to follow the plate map and aliquot the samples using the multichannel pipette, aliquot the samples in the strip tubes in replicates to mimic the plate map.
6. Using correct volume-range pipettes for greater accuracy, prepare a large master mix without the primers and aliquot the appropriate amount into each of the nine strip tube wells.
7. Add the indicated amount of each of the nine primer mix in each of the nine strip.
8. After aliquoting the master mix in the strip tubes and adding well-specific primers to each of the strip tube wells, cap the strip tubes, vortex and centrifuge and, using multichannel automated pipette, transfer 3.0 μl of the Master Mix from the 0.2 ml PCR reaction tubes to the 96-well plate according to the plate map.
9. Add 2 μl (10 ng/μl) of DNA to the appropriate wells, using a cap opener to open the capped strip tubes to avoid cross contamination.
10. Seal the sample plate with a plastic sealer with the help of a microseal film roller, briefly vortex the 96-well PCR reaction plate, remove air bubbles, and spin down in Pre-PCR plate centrifuge at 4000 rpm.

3.2 PCR Conditions-Sequenom-PCR

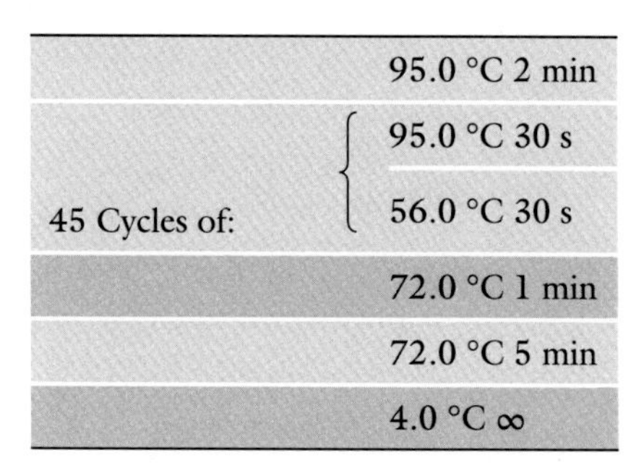

	95.0 °C 2 min
45 Cycles of:	95.0 °C 30 s
	56.0 °C 30 s
	72.0 °C 1 min
	72.0 °C 5 min
	4.0 °C ∞

After 2.5 h, the amplification is completed. Proceed to the SAP reaction step or store the plate at 4 °C.

3.3 Post-PCR SAP Reaction Cleanup

Shrimp alkaline phosphatase is used to remove excessive unincorporated quantities of dNTPs from amplification products, a process that is achieved by cleaving the phosphate groups from the 5′ termini [5, 11].

The following steps should be followed in order to dephosphorylate unincorporated dNTPs:

1. After the PCR reaction is completed, the PCR product is treated with Shrimp Alkaline Phosphatase (SAP) to neutralize unincorporated dNTPs (refer to principle above).
2. SAP treatment is done on the POST-PCR work area. SAP buffer is allowed to thaw at room temperature for 10 min.
3. The reaction box for SAP step should be already filled in based on the number added to the first PCR step.
4. Aliquot the indicated amount from the large master mix to 8 strip well tubes.
5. PCR plate should be quick spun for 15 s at 4000 rpm in the Post-PCR plate centrifuge and the microseal is removed.
6. Add 2.0 μl of the SAP reaction in each of the 96 wells of the PCR plate containing 5 μl PCR product using automatic 8-channel pipette.
7. The final volume should now be 7.0 μl in each of the wells of the 96-well PCR plate.
8. Seal the sample plate with microseal using plate roller, vortex to mix, remove bubbles, and centrifuge the plate in the POST-PCR centrifuge at 4000 rpm for 15 s.
9. Verify that the plate is properly labeled with run number, date, tech initials, and the SAP step.
10. Place the PCR reaction plate in ABI 2720 thermal cycler following the same PCR plate number and select the following Sequenom SAP program.

11. PCR conditions for SAP.

    37 °C 40 min.
    85 °C 5 min.
    4 °C ∞.

12. SAP reaction is completed in 45 min. Place the plate at 4 °C if performing extension primer adjustments or proceed to extension reaction setup.

3.4 Single-Base Primer Extension, iPLEX pro Reaction

For the iPLEX assay all reactions are terminated after a single-base extension [11]. Here a primer extension master mix (which contains extension primers, buffer, enzyme, and the mass-modified ddNTPs) is added to the amplification products [5]. The mass-modified terminators incorporated in the iPLEX assay were designed to overcome the restrictive nature of small mass separation of single-base extension in high-throughput genotyping [5, 11].

Note: *Extension reaction is setup on the Post-PCR room.*

1. Thaw and mix all extension reaction reagents (except for enzyme and termination mix) before setting up the extension reaction plate.
2. As already mentioned, iPLEX Pro cocktail containing extension primers, buffer, enzyme, and terminator nucleotide is used for single-base primer extension reaction just before the SNP site.
3. As the reagents thaw, spin down the SAP plate and note all the lot numbers on the worksheet.
4. As with the PCR, Extension master mix is first created in a large tube and then aliquoted into nine strip tube wells as indicated in the worksheet. The number inside the number of reaction box should be automatically filled and should reflect the number of samples being run.
5. Aliquot the indicated amount of the master mix in each of the nine strip tube wells.
6. To each of the nine strip tube wells with the extension master mix, add the indicated amount of each of the nine extension primers.
7. Remove the plate from the thermal cycler, spin down in the Post-PCR plate centrifuge at 4000 rpm for 15 s.
8. Using multichannel automated pipette, transfer 2.0 μl of each of the nine master mixes to the appropriate wells of the 96-well plate with 7 μl of Post-PCR and SAP mixture to bring the final volume to 9 μl.
9. Seal plate, vortex lightly, remove air bubbles and centrifuge briefly at 4000 rpm for 15 s.
10. Place the reaction plate in ABI 2720 thermal cycler following the same plate order as SAP and select the following Sequenom Extension reaction program.

11. Extension Reaction.

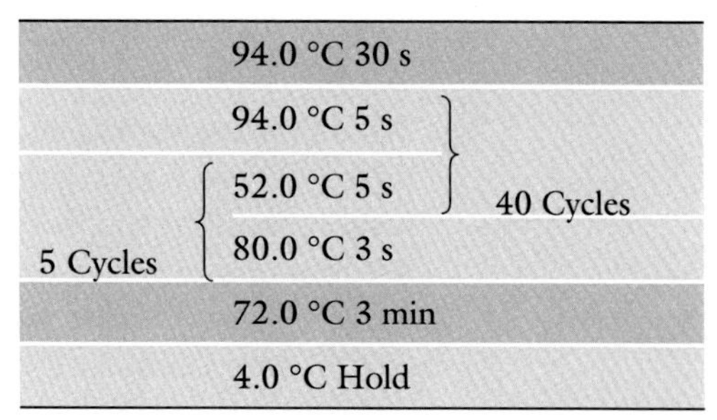

	94.0 °C 30 s	
	94.0 °C 5 s	40 Cycles
5 Cycles	52.0 °C 5 s	40 Cycles
5 Cycles	80.0 °C 3 s	
	72.0 °C 3 min	
	4.0 °C Hold	

12. Extension reaction is completed overnight. Sample dispensation, mass spectrometry, and data analysis should be done the following day.

3.5 Volume Checks and Sample Dispensation

1. The final extension product is dispensed from the skirted 96-well plate into a Spectro Chip designed for 96-well format through the MassArray Nanodispenser unit. The unit consists of two sets of six pins, which aspirates the sample from the plate and dispenses nanoliters of the sample on to the chip.
2. After the extension reaction, the 96-well skirted plate is spun down for 15 s at 4000 rpm (program 5 of the post-PCR plate-centrifuge) to collect the droplets from the side of the plate.
3. Remove Three-Point Calibrant from −20° freezer and allow it to thaw at room temperature.
4. Obtain 96-well dimpled plate and spread clean resin onto the plate.
5. Ensure that all dimpled wells are covered evenly with the resin, using extra flat utensil provided to push the resin into the wells.
6. Unseal the plate and add 41 μl of HPLC water into each of the 96 wells of the sample plate to bring the final volume up to 50 μl.
7. Invert the plate onto the dimpled plate with resins, ensuring that the wells in the sample plate line up with the wells of the resin plate.
8. Invert the plate again to allow resin to fall into the sample plate.
9. Ensure that all the resin from the dimpled well has been dispensed into each of the sample well.
10. Seal the plate, label with run number, date, and tech initials and allow the plate to rotate on a sample rotator for at least 30 min.
11. While the plate is rotating, program your plates into the Spectro Typer 4.0 software.

12. After rotating the plate with sample and clean resin, spin the plate at 3200×g for 5 min (Program 9 of the post-PCR plate centrifuge).
13. The centrifugation will allow the liquid in the wells to be at least 2.5 mm above the recommended level of the resin. The resin remains at the bottom of the well with the purified product remaining in the supernatant ready to be removed by the SpectroPoint [5]. At this point plates may be stored at 4 °C (stable for around a month) or may be ready to be spotted into the SpectroCHIPs [5].
14. The instrument has a number of predefined methods that can be selected or new methods can be created based on the number of samples, the location of samples on the well, and where the samples are to be dispensed on the array chip.
15. After centrifuging the plate, open the main door of the nanodispenser to disengage the safety lock (dialog box appears).
16. Place the microtiter plate (MTP) on the plate holder 1 and orient the plate so that well A1 is to the lower left, facing forward.
17. Leave the main door open to load the SpectroCHIPs into a standard scout plate adapter.
18. The standard scout plate adapter can hold up to 10 SpectroCHIPs together, but only position 1 and position 2 are used.
19. Insert SpectroCHIPs into the first position of the scout plate using tweezers or forceps to pick up the chip by the edges from the plastic chip holder, being careful not to touch the surface of the SpectroCHIPs.
20. Open tab labeled ASPIRATE/DISPENSE at the bottom of the METHOD screen.
21. Select ANALYTE ONLY under OPERATION, WET RUN under SPOTTING, adjust ASPIRATE SETTINGS and DISPENSE SETTINGS to the following:
    (a) ASPIRATE TIME: 8 s.
    (b) ASPIRATE OFFSET: 6.5 mm.
    (c) ASPIRATE SPEED: 60 mm/s.
    (d) DISPENSE TIME: 1 s.
    (e) DISPENSE OFFSET: 2 mm.
    (f) DISPENSE SPEED: 220 mm/s.
22. After adjusting all the parameters, press BACK icon on the top right of the screen and press APPLY. On the MAIN MENU, select TRANSFER and click on METHOD icon.
23. With the sample plate and a SpectroCHIP appropriately placed within the dispenser, select STEP on the transfer screen to perform one-step volume check.

24. The 6-pin array fixture will align and position itself to the SpectroCHIP. After the alignment, the run begins. Press VOLUME icon on the top to check the volume dispensed on the chip.
25. Check the volume of each of the pins and look at the mean volume and standard deviation given at the bottom of the screen. Tap BACK button after checking the volume to return to the TRANSFER screen.
26. If the dispensation speed gives desired volume, park the pins and replace the current SpectroCHIP used for volume check with a new chip for running the samples.
27. Replace the chip and fill the calibrant reservoir, located in the dry station, with 65 μl of three-point calibrant. On the MAIN MENU, select MAPPING to map out a plate to a chip and select RUN arrow on the TRANSFER screen to start a full plate run.
28. After rinsing and purging lines and alignment of the camera and the pin array with the chip, run will begin. It takes approximately 15–20 min to complete dispensing from one 96-well microtiter plate to one 96 SpectroCHIP using standard 6-pin array format.

3.6 Detection by Mass Spectrometry

1. Double click on RT PROCESS icon to start all RT process containing Flex Control, Server Control, Caller and Acquire. Once the RT process has been started, maximize the FlexControl and SpectroAcquire programs. On the SpectroAcquire window, click on AUTO RUN SETUP tab.
2. In the blank space labeled as CHIP 1, type in the SpectroCHIP barcode.

   The barcode name is the same name as used to link the chip. Press BARCODE REPORT button. A popup window should show that the chip is FOUND and it should be in green.
3. Before the nanodispenser finishes dispensing the last three steps transfer of the 96-Well MTP on the SpectroCHIP, press the TARGET IN/OUT button on the Mass Spectrometer control panel to extend the target.
4. Open the lid to obtain the chip holder. Close the lid and place the SpectroCHIP with the sample dispensed from the nanodispenser to the chip holder in the correct position.
5. Place the chip holder back into the instrument, close the lid, and press TARGET IN/OUT button to retract the target.
6. Wait for the READY light to illuminate on the instrument control panel. The ready light indicates that the instrument is ready for an Auto Run. It takes approximately 2 min for the target to be fully retracted.

7. Once the ready light is illuminated, bring up the FlexControl window.
8. Under SPECTROMETER tab, turn ON the HIGH VOLTAGE. On the AUTORUN tab of the SpectroAcquire window, press START AUTO RUN. The Chip ID will be visible and the run will begin.

3.7 Mass Array Spectrometer Analysis and Interpretation

Genotype calls are done in a real time fashion during the chip detection and traces are available right after detection using the software tools [5].

1. After the end of the Auto Run, the data for the chip of interest will be automatically saved in Typer Analyzer under the sample name.
2. To access your chip, bring up the Typer 4.0 tool bar and click on Typer Analyzer.
3. On the active chip, click the box to select the CHIP to be analyzed.
4. The traffic light panel gives a quick assessment of percentage (%) successful call rate of an assay per well. It ranges from dark green color (best) to light green, yellow and red being no call or bad spectra.
5. On the very top, click on VIEW and DETAILS pane to reveal a new window showing spectra. Use Detail Panel to look at each SNP in each of the nine wells for each of the samples and match with the spectra.

Conservative calls are the most confident calls, indicating that all the extension primer has been used. Moderate calls are usually a result of excess noise and some unused extension primer seen as a small peak. Aggressive calls are an indication of leftover excess extension primers due to low PCR product. Low probability calls are due to heterozygous mutations and/or missing or misaligned peaks and warrants manual inspection and interpretation. No alleles and No calls are due to bad spectrum caused by lack of PCR and/or extension product or due to bad spotting by nanodispenser.

4 Notes

1. Adjustment of extension primer is a crucial step in the protocol (Subheading 2.3); as with increasing mass, signal-to-noise ratios tend to decrease [11]. In some instances signals may become indistinguishable from noise leading to calling errors. Extension primers are adjusted to ensure equal Signal-To-Noise Ratio (SNR) of all the primers found in each of the wells. Adjustments are done for any and all the new extension primer mix before performing the extension reaction.

2. Primer extension reaction resin cleanup (Subheading 3.5, **step 4**) is a crucial step in order to desalt the iPLEX reaction products and thus optimize mass spectrometry analyses [5, 11]. This is achieved by using a cationic resin pretreated with acid reagents (SpectroClean from Sequenom) which removes salts such as Na+, K+, and Mg2+ ions. Removal of these ions avoids generation of high background noise in the mass spectra [5].
3. While loading the chip for mass spectroscopy (Subheading 3.6, **step 4**), it is recommended that the chip holder is taken inside the dispenser and transfer the newly dispensed sample chip to the chip holder under the dispenser to avoid exposing the chip to air and any possible accidents.

References

1. Vogelstein B, Papadopoulos N, Velculescu VE, Zhou S, Diaz LA Jr, Kinzler KW (2013) Cancer genome landscapes. Science 339:1465–1558
2. Harris TJ, McCormick F (2010) The molecular pathology of cancer. Nat Rev Clin Oncol 7(5):251–265
3. Klein S, McCirmick F, Levitzki A (2005) Killing time for cancer cells. Nat Rev Cancer 5(7):573–850
4. Dar AC, Shokat KM (2011) The evolution of protein kinase inhibitors from antagonists to agonists of cellular signaling. Annu Rev Biochem 80:769–795
5. Gabriel S, Ziaugra L, Tabbaa D (2009) SNP genotyping using the Sequenom MassARRAY iPLEX platform. Curr Protoc Hum Genet. Chapter 2:Unit 2.12
6. Tang K, Fu DJ, Julien D, Braun A, Cantor CR, Koster H (1999) Chip-based genotyping by mass spectrometry. Proc Natl Acad Sci U S A 96(18):10016–10020
7. Karas M, Bachman D, Hillenkamp F (1985) Influence of the wavelength in high-irradiance ultraviolet laser desorption mass spectrometry of organic molecules. Anal Chem 57(14): 2935–2939
8. Fenn JB, Mann M, Meng CK, Wong SF, Whitehouse CM (1989) Electrospray ionization for mass spectrometry of large biomolecules. Science 246(4926): 64–71
9. Karas M, Hillenkamp F (1988) Laser desorption ionization of proteins with molecular masses exceeding 10.000 daltons. Anal Chem 60(20):2299–2301
10. Honisch C, Chen Y, Hillenkamp F (2010) Comparative DNA sequence analysis and typing using mass spectrometry. In: Shah HN, Gharbia SE (eds) Mass spectrometry for microbial proteomics, vol 19. Wiley, Chichester, pp 443–462
11. Oeth P, Beaulieu M, Park C, Kosman D, del Mistro G, van den Boom G, Jurinke C (2005) iPLEX™ assay: increased plexing efficiency and flexibility for MassARRAY® system through single base primer extension with mass-modified terminators. SEQUENOM® Application note. Doc No. 8876–006, R01. CO 050154

Chapter 10

Microfluidics-Based PCR for Fusion Transcript Detection

Hui Chen

Abstract

The microfluidic technology allows the production of network of submillimeter-size fluidic channels and reservoirs in a variety of material systems. The microfluidic-based polymerase chain reaction (PCR) allows automated multiplexing of multiple samples and multiple assays simultaneously within a network of microfluidic channels and chambers that are co-ordinated in controlled fashion by the valves. The individual PCR reaction is performed in nanoliter volume, which allows testing on samples with limited DNA and RNA. The microfluidics devices are used in various types of PCR such as digital PCR and single molecular emulsion PCR for genotyping, gene expression, and miRNA expression. In this chapter, the use of a microfluidics-based PCR for simultaneous screening of 14 known fusion transcripts in patients with leukemia is described.

Key words Integrated fluidic circuits, Dynamic array, Microfluidics-based PCR, Real-time PCR

1 Introduction

Microfluidics technologies integrate advances from multidisciplinary fields intersecting engineering, physics, chemistry, biochemistry, nanotechnology, and biotechnology to design systems which have precise control and manipulation of fluids that are geometrically constrained to a small, typically submillimeter scale [1–3]. The microfluidics technology allows use of a wide variety of sample types and multiple chemistry choices with flexibility in sample manipulations. The application of microfluidics technologies in molecular diagnostics has revolutionized the field by allowing high-throughput and multiplexed diagnostic screening for bacterial detection, viral detection, cancer, as well as non-cancer biomarker detection [2, 4–6]. The microfluidics devices can be used for high-throughput multiplex PCR, real-time PCR, reverse transcriptase PCR, digital PCR, and single molecular emulsion PCR [7, 8]. The individual PCR reaction is performed in nanoliter (nL) volume, which allows testing on samples with limited DNA and RNA.

Rajyalakshmi Luthra et al. (eds.), *Clinical Applications of PCR*, Methods in Molecular Biology, vol. 1392,
DOI 10.1007/978-1-4939-3360-0_10, © Springer Science+Business Media New York 2016

This chapter focuses on Fluidigm Dynamic Array (Biomark) designed for high-throughput real-time PCR for cancer biomarker testing focusing on leukemia. The microfluidics-based PCR devises such as Fluidigm integrated fluidic circuits (IFCs) integrate thermal cycling and fluorescence detection on a single array/chip. The Dynamic Array IFCs have an on-chip network of microfluidic channels, chambers, and valves that automatically assemble individual PCR reactions and significantly decreasing the number of pipetting steps required. The loading of individual patient samples to sample chambers (9 nL each for 48.48 dynamic array) is preceded by the closure of interface valves and opening of containment valves, and followed by the loading of assay mixtures containing primers and probes to assay chambers (1 nL each for 48.48 dynamic array). Then the containment valves are closed and interface valves are open to allow pair-wise mixing procedure of thousands of individual PCR reactions. Fluidigm Dynamic IFC Array can be used for screening the known clinically relevant translocations in leukemia, lymphoma, sarcoma, carcinoma, and other tumors. Screening currently for known leukemia-related fusion transcripts in multiple leukemia patients can be accomplished on a single Fluidigm Dynamic Array kit using reverse transcriptase PCR and Taqman real-time PCR within few hours.

In particular, the protocol describes a high-throughput multiplex reverse transcriptase Taqman PCR based clinical assay using Fluidigm Dynamic Array IFCs for screening 14 known fusion transcripts in leukemia. The classification of distinct subtypes of hematopoietic neoplasms is defined in part by specific gene rearrangements due to chromosomal translocations [9]. The presence or absence of certain disease-associated translocations predicts response to targeted therapy and clinical outcome. For example tyrosine kinase inhibitor, Imatinib, is used for treating patients with chronic myeloid leukemia and B-lymphoblastic leukemia with BCR-ABL fusion gene resulting from t(9;22) and all-trans retinoic acid is used for treating patients with acute myeloid leukemia with t(15;17). Thus testing for the presence of these recurrent translocations at diagnosis for accurate diagnostic subclassification and therapeutic and prognostic stratification has become standard of care. The protocol described here is used for newly diagnosed leukemia cases to screen 14 known fusion gene transcripts that include *RUNX1-RUNX1T1*, *CBFβ-MYH11* variants A, D, and E, *PML-RARA* long, short, and alternative forms, *BCR-ABL1* b2a2, b3a2, and e1a2 variants, *ETV6-RUNX1*, *E2A-PBX1*, *MLL-AF4*, and *DEK-CAN*. Compared to orthogonal singleplex methods which require multiple setups of individual reactions to cover initial screening of new patients, the microfluidics-based PCR would dramatically reduce the sample volume requirement, testing time, technologist's hand-on time, and overall cost.

2 Materials

2.1 Sample Requirements

Bone marrow or peripheral blood specimen collected in EDTA (purple/lavender—top tube) shipped on wet ice.

2.2 Equipment and Supplies

BioMark HD system (Fluidigm).
IFC controller (Fluidigm).
Dynamic Array (Fluidigm).
Laminar flow hood.
Picofuge.
96 Well PCR plates.
Eppendorf tubes.
Pipettors and tips.
Positive and negative control cell lines (NB4, KBM7, KASUMI, K562, and HL60).

2.3 Generic Reagents

2× TaqMan PreAmp master mix (Applied Biosystems).
2× Taqman universal PCR master mix (Applied Biosystems).
25× Ipsogen gene expression assay (Qiagen).
20× TaqMan® gene expression assays (Applied Biosystems).
2× Assay loading reagent (Fluidigm).
20× GE sample loading reagent (Fluidigm).
Superscript II RT kit (Invitrogen).
10 mM Tris, pH 8.0, 0.1 mM EDTA.
Deionized DNA-free, DNase-free, RNase-free water.

2.4 Primers and Probes

Forward primers, reverse primers, and Taqman probes for the following fusion genes and control:
RUNX1-RUNX1T1.
CBFβ-MYH11 variant A, variant D, and variant E.
PML-RARA long form, short form, and alternative form.
BCR-ABL1 b2a2, b3a2, and e1a2.
ETV6-RUNX1.
E2A-PBX1.
MLL-AF4.
DEK-CAN.
ABL (control).

2.5 Reagent and Equipment QC and Maintenance

2.5.1 Reagent QC

The reagents are first tested on negative and the positive controls. If satisfactory results are obtained, the reagents are then used on patient samples.

2.5.2 Equipment QC

BioMark™ equipment Maintenance and Calibration are performed by BioMark™ once a year.

2.5.3 Calibration

Calibration is performed with proper controls and standards within each run. The criteria for accepting a valid run include the cycle threshold (Ct) for a pooled 5 point sensitivity control that is below or equal to the sensitivity cutoff derived during test validation and Ct for negative and reagent controls greater than the assay cutoff set during test validation.

3 Methods

RNA is extracted from blood or bone marrow following standard RNA extraction procedure in designated pre-PCR area. Reverse transcription is performed to convert mRNA in patient samples to cDNA using Superscript II RT kit with 0.1 μg/μL of RNA (*See* **Note** in heading 4 for general PCR contamination precautions). The cDNA can be stored at 2–6 °C for short term and at −15 to −25 °C for long-term storage.

3.1 Preamplification

As very small amounts of input cDNA is used in the assay using Dynamic Array IFC PCR, to obtain the best qualitative and quantitative result preamplification of cDNA (PreAmp) before real-time PCR is recommended.

(a) Prepare 0.2× pooled assay mix by mixing 1 μL of 20× gene expression assay and 99 μL of DNA suspension buffer per assay with appropriate scale-up for multiple reactions in the run.

(b) Prepare preamplification reaction according to reagents and volumes in Table 1. First prepare the stock solution containing 2× PreAmp master mix and 0.2× pooled assay mix. The volume requirement for each reaction is 2.5 μL of 2× PreAmp master mix and 1.25 μL of 0.2× pooled assay mix. Mix the stock solution. Then aliquot 3.75 μL of the stock solution to the individual well of PCR plate. Add 1.25 μL of cDNA to each well. Mix the components by pipetting up and down twice.

Table 1
Preparation for preamplification reaction

			Reactions	
Reagents	**Lot #**	**Expiration date**	**1**	**60 (Stock)**
PreAmp master mix (2×) (μL)			2.5	150
Pooled assay mix (2×) (μL)			1.25	75
Subtotal volume (μL)			3.75	225
cDNA (μL)			1.25	
Volume/well (μL)			5	

(c) Place PCR plate in a thermocycler. Set initial denaturation at 95 °C for 10 min, followed by 14 cycles of 95 °C for 15 s and 60 °C for 4 min. The temperature at the end of PreAmp PCR is set to 4 °C.

(d) The final PreAmp PCR products (5 μL each) are diluted 1:5 with 20 μL DNA suspension buffer for each reaction in order to dilute any remaining PCR by-products.

3.2 Prime the Dynamic Array IFC

(a) Inject the control line fluid into the accumulators on both sides of the chip using syringes recommended for corresponding array chip (300 μL control line fluid for 48.48 Dynamic Array IFC).

(b) Remove the blue protective film from the bottom of the chip.

(c) Place the chip into the IFC Controller MX, then run the corresponding Prime script to prime the control line fluid into the chip.

(d) The chip needs to be used within certain time frame after the package is open, within 24 h for 48.48 dynamic array chip. Load the chip within 60 min of priming.

3.3 Prepare Analyte 10× Assay Solution for Assay Inlets

(a) Calculate the stock solution volumes for analyte 10× assays with appropriate scale-up for multiple reactions in the run according to the reagents and volumes listed in Table 2. Analyte 10× assay solution for single reaction contains 2.5 μL of 20× TaqMan gene expression assay and 2.5 μL of 2× assay loading reagent. The final concentration at 10× is 9 μM for primers and 2 μM for probe.

(b) In a DNA-free hood, prepare and aliquot the stock for analyte 10× assay solution.

3.4 Prepare Sample Pre-mix Solution and Sample for Sample Inlets

(a) Calculate the stock solution volumes for sample pre-mix with appropriate scale-up to accommodate multiple runs according to the components and volumes listed in Table 3. Sample pre-mix for single reaction contains 2.5 μL of 2× TaqMan universal PCR master mix and 0.25 μL of 20× GE sample loading reagent.

Table 2

Analyte 10× assay solution for single reaction and stock solution for 40 reactions

Reagents	Lot #	Expiration date	Reactions 1	Overage	40 (Stock)
20× TaqMan® gene expression assay (μL)			2.5	3	100
2× Assay loading reagent (μL)			2.5	3	100
Total volume (μL)				6	200
Volume/inlet (μL)			5		

Table 3
Sample pre-mix solution for single reaction and stock solution for 60 reactions

Reagents		Lot #	Expiration date	Reactions		
				1	Overage	60 (Stock)
Pre-mix	TaqMan® Universal PCR MM (2×) (μL)			2.5	3	180
	20× GE sample loading reagent (μL)			0.25	0.3	18
Diluted PreAmp product (μL)				2.25	2.7	
Total volume (μL)					6	198
Volume/inlet (μL)				5		

(b) In a DNA-free hood, prepare sample pre-mix stock solution in a 1.5 mL sterile tube.

(c) Aliquot 3.3 μL of the sample pre-mix for each sample reaction in the DNA-free hood.

(d) Remove the aliquots from the DNA-free hood and add 2.7 μL of diluted PreAmp PCR product to each well to make a total volume of 6 μL in each aliquot of sample solution.

3.5 Loading the Chip

(a) At the end of chip priming, remove the primed chip from the IFC Controller MX.

(b) Vortex all the assay and sample solutions thoroughly to mix the components and centrifuge the solution to the bottom of microfuge tubes or microplate wells. Pipette 5 μL of each assay solution from Subheading 3.3 into assay inlets on the left side of the primed chip and 5 μL of each sample solution from Subheading 3.4 into sample inlets on the right side of the primed chip (Fig. 1).

(c) The unused sample inlets need to be filled with 3.3 μL of sample pre-mix and 2.7 μL of DNA-free water per inlet. The unused assay inlets need to be filled with 3.0 μL assay loading reagent and 3.0 μL of water. Pipette no template control (NTC) into sample inlet #22 for 48.48 dynamic array chip.

(d) Place the chip to the IFC Controller MX. Run the load mix script in IFC Controller MX software to load the samples and assays into the chip.

(e) Remove loaded chip from the IFC Controller MX after the load mix script has finished. Remove any dust particles or debris from the chip surface to prevent interference of signal detection during data collection. Continue to run loaded chip on microfluidics real-time PCR instrument within 4 h of loading the chip.

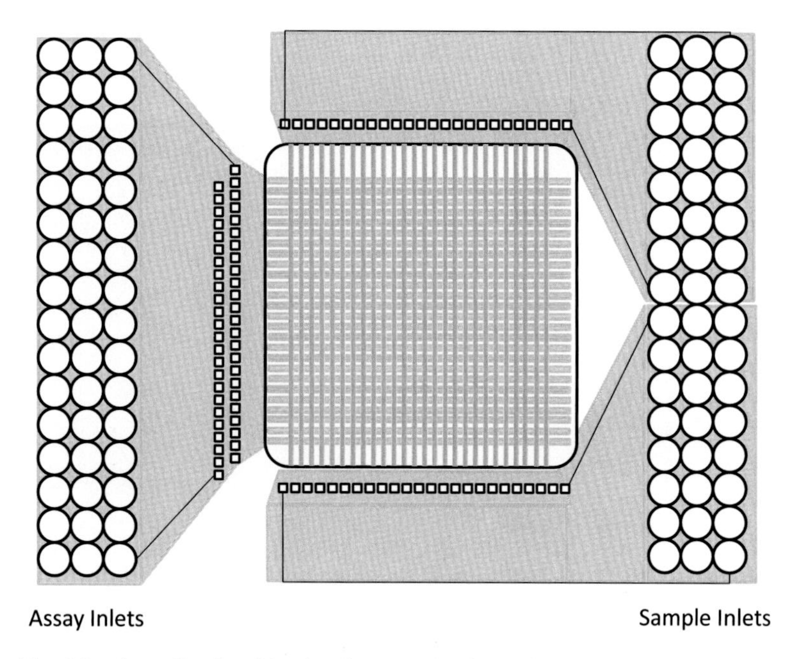

Fig. 1 A schematic of a chip pipetting map for dynamic array IFC

3.6 Data Collection

(a) Start a new run of microfluidics real-time PCR using the data collection software for BioMark HD system. Check the status bar to verify that the lamp and the camera are ready (green) before proceeding to the run.

(b) Place the loaded chip into the reader and click load to load the chip in BioMark HD system. Verify chip barcode and chip type for the run. Select project settings, file location for data storage, type of assay (i.e., single probe, two probes, or more than two probes), and probe types.

(c) Choose the appropriate thermal cycling protocol file, GE 48×48 Standard v1.pcl for 48.48 dynamic array. Confirm auto-exposure for real-time PCR. After final verification of the chip run information and click start run to start microfluidics real-time PCR.

3.7 Data Analysis and Result Interpretation

Data analysis is performed using the Fluidigm Real-Time PCR Analysis software to automatically determine the cycle threshold and to calculate the standard curve for each chip run. For screening leukemia translocation panel at the patient samples, standards and controls are run in triplicates. An acceptable run should have threshold cycle (Ct) less than 25 for a pooled 5 point sensitivity control and positive controls and Ct greater than 30 for negative and reagent controls. A positive result is called if Ct is less than 25 in triplet chambers (Figs. 2 and 3). ABL mRNA is a ubiquitous transcript present in all nucleated cells and serves as RNA quality

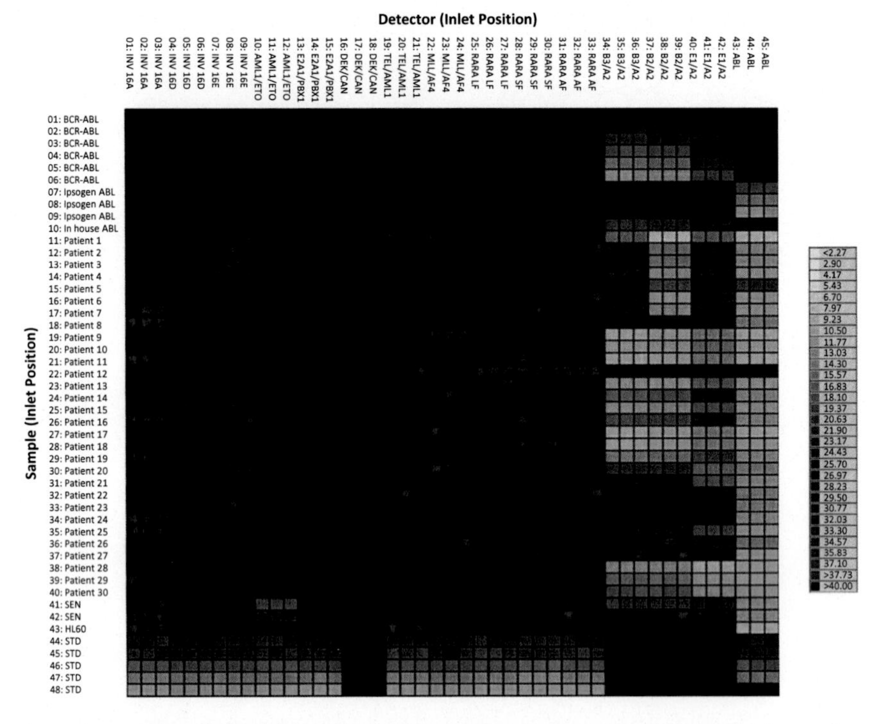

Fig. 2 Heatmap Image of a 48.48 chip run for leukemia translocation panel. The samples and controls are run in triplicates. The color codes at the bottom indicate the corresponding PCR cycles to detect PCR product. The chamber is coded black if no amplification after 40 cycles of real-time PCR

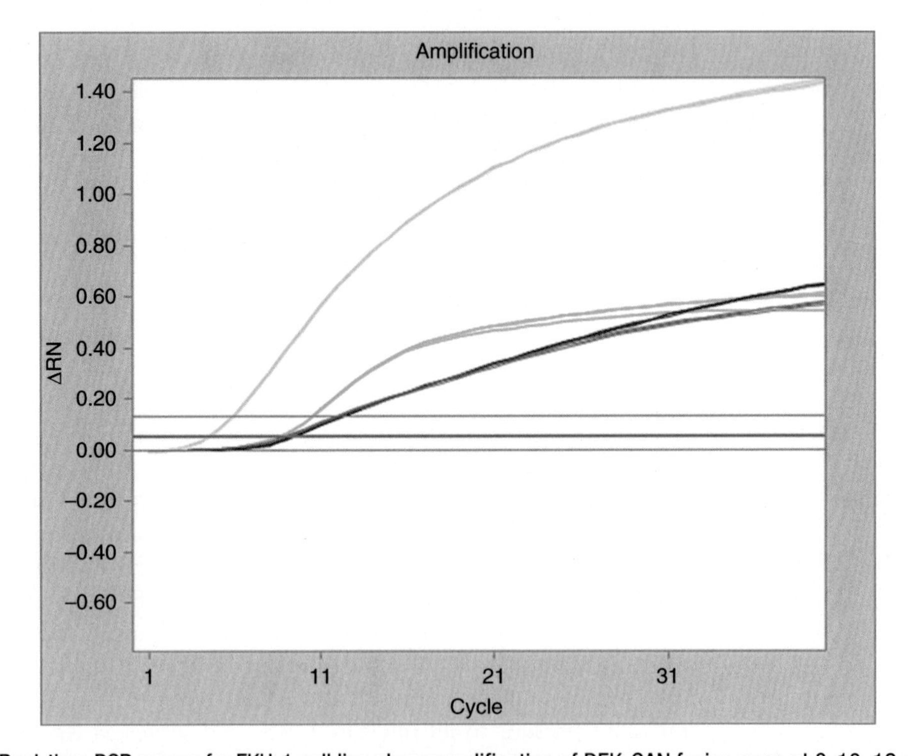

Fig. 3 Real-time PCR curves for FKH-1 cell line show amplification of DEK-CAN fusion gene at 6, 10, 12 cycles

control for the sample. A negative result (undetectable) is valid if (1) ABL copy number is higher than the copy number set by the design of assay (Ct between 13.4 and 19.4 for screening leukemia translocation panel); and (2) no signals or weak signals are seen but not meet the criteria for positive results (Ct > 30).

4 Notes

PCR Contamination Precautions: The general contamination precautions must be taken at every single moment during the entire experiment since the sensitivity of real-time PCR is near single copy detection. The PCR-related activities are physically divided into two separate rooms, pre-PCR room and PCR room. A unidirectional workflow from pre- to post-PCR must be maintained at all times. Pre-PCR room is designated for nucleic acid extraction and PCR setup. Samples and reagents preparation should be performed in the UV-sterilized PCR preparation hood in pre-PCR room. Prior to and after each single use, the UV light is left on for a half-hour in the PCR preparation hood in order to decontaminate any traces of nucleic acids by cross-linking. Post-PCR room is used for PCR and after PCR activities.

Acknowledgements

I would like to thank Dr. Rajyalakshmi Luthra, Dr. Rajesh Singh, Seema Hai, and molecular diagnostic laboratory at MD Anderson Cancer Center for providing figures and technical support.

References

1. Kulinsky L, Noroozi Z, Madou M (2013) Present technology and future trends in point-of-care microfluidic diagnostics. Methods Mol Biol 949:3–23
2. Jayamohan H, Sant HJ, Gale BK (2013) Applications of microfluidics for molecular diagnostics. Methods Mol Biol 949:305–334
3. Moltzahn F et al (2011) High throughput microRNA profiling: optimized multiplex qRT-PCR at nanoliter scale on the fluidigm dynamic arrayTM IFCs. J Vis Exp (54). pii:2552
4. Kim MS, Kwon S, Park JK (2013) Breast cancer diagnostics using microfluidic multiplexed immunohistochemistry. Methods Mol Biol 949:349–364
5. Wang J et al (2009) High-throughput single nucleotide polymorphism genotyping using nanofluidic dynamic arrays. BMC Genomics 10:561
6. Dhoubhadel BG et al (2014) A novel high-throughput method for molecular serotyping and serotype-specific quantification of Streptococcus pneumoniae using a nanofluidic real-time PCR system. J Med Microbiol 63(Pt 4):528–539
7. Jang JS et al (2011) Quantitative miRNA expression analysis using fluidigm microfluidics dynamic arrays. BMC Genomics 12:144
8. Leng X, Yang CJ (2013) Agarose droplet microfluidics for highly parallel and efficient single molecule emulsion PCR. Methods Mol Biol 949:413–422
9. Swerdlow SH, Campo E et al (2008) WHO classification of tumours of haematopoietic and lymphoid tissues, 4th edn. WHO press, Lyon

Chapter 11

Polymerase Chain Reaction Diagnosis of Leishmaniasis: A Species-Specific Approach

Eglys González-Marcano, Hirotomo Kato, Juan Luis Concepción, María Elizabeth Márquez, and Alberto Paniz Mondolfi

Abstract

Leishmaniasis is an infectious disease caused by protozoan parasites of the genus *Leishmania* which are transmitted to humans through bites of infected sand flies. The variable clinical manifestations and the evolution of the disease are determined by the infecting species. Recognition at a species level is of utmost importance since this greatly impacts therapy decision making as well as predicts outcome for the disease. This chapter describes the application of polymerase chain reaction (PCR) in the detection of *Leishmania* parasites across the disease spectrum, including protocols for sample collection and transportation, genomic material extraction, and target amplification methods with special emphasis on PCR amplification of the cytochrome *b* gene for *Leishmania spp.* species identification.

Key words Leishmaniasis, Diagnosis, PCR

1 Introduction

The standard methods for diagnosis of cutaneous and visceral Leishmaniasis have traditionally relied on the direct identification of amastigotes by histology, direct microscopy (Giemsa-stained smears), or by the growth of promastigotes in axenic culture [1] and/or laboratory animals. Recent clinical and laboratory research trials have confirmed that not only the immune response of the host but also the species of infecting parasite are responsible in determining the clinical manifestations of leishmaniasis and affect the response to treatment [2, 3]. In this sense, polymerase chain reaction (PCR) has emerged as a powerful tool for diagnosis of this parasitic disease by providing a higher sensitivity, the ability to detect low levels of parasitemia, and the capacity of detecting mixed infections. Most importantly, PCR allows accurate molecular characterization of the causative agent, which is essential to evaluate the efficacy of anti-leishmanial drugs against different circulating strains. In the following chapter, we discuss the general aspects of DNA

Rajyalakshmi Luthra et al. (eds.), *Clinical Applications of PCR*, Methods in Molecular Biology, vol. 1392,
DOI 10.1007/978-1-4939-3360-0_11,

extraction methods, tissue processing, and the wide array of PCR leishmanial targets with special emphasis on the PCR amplification of the cytochrome *b* gene for *Leishmania spp.* species identification [4, 5]. This protocol offers the advantage of collecting and extracting the samples from FTA Classic Cards, making it an ideal and practical method to work with on the field while offering a robust target for identification and phylogenetics of *Leishmania* parasites.

2 Materials

2.1 Equipment

1. Safety cabinet.
2. Water bath.
3. Centrifuge/Micro-centrifuge.
4. Vortex.
5. Pipettes for PCR.
6. Filter tips.
7. DNA quantifier equipment.
8. Thermocycler.
9. Agarose gel electrophoresis equipment.
10. Sequence Detection System (real-time PCR).

2.2 Reagents

1. AxyPrep™ Blood Genomic DNA Kit (Axygen).
2. Phenol:Chloroform:Isoamyl Alcohol (Invitrogen).
3. PCR kit.
4. Primers.
5. Nuclease-free water.
6. Agarose.
7. SYBR® Safe DNA Gel Stain/Ethidium bromide.
8. Restriction enzymes and buffers.
9. FTA Classic Cards.
10. Ampdirect Plus reagent (Shimadzu Biotech, Tsukuba, Japan).

3 Methods

Herein we describe a number of protocols for the diagnosis of visceral (VL), cutaneous (CL), and mucocutaneous leishmaniasis (MCL); therefore, it is important to highlight the differences and adequate type of sample to be used for retrieving DNA templates in each case scenario. For VL the most commonly used sample is whole blood, sometimes used in combination with buffy coat in order to obtain a better yield of DNA concentration. However bone marrow or/and spleen aspirates are also acceptable specimens.

For CL and MCL samples used vary from tissue aspirates and skin scrapings taken from the active lesion to skin biopsies.

3.1 Sample Selection and Storage

Cutaneous (CL) or Mucocutaneous Leishmaniasis (MCL) should be rapidly treated to avoid the potentially disfiguring consequences of parasite spread. Visceral Leishmaniasis (VL), on the other hand, must be promptly diagnosed and treated due to its fulminating and usually rapidly fatal outcome. Therefore early and reliable methods are needed to diagnose this important spectrum of parasitic diseases.

In recent years, molecular techniques for the detection of leishmanial DNA or RNA have emerged as a promising resource to be used not only in diagnosis but also in species identification while providing a higher sensitivity than microscopy and culture. In this context, sample preparation and DNA extraction methods can greatly influence the outcome and reliability of the PCR. In this type of infectious diseases, this process is complicated due to the abundance of host DNA in the samples, which can compete with the parasites' DNA (generally much less abundant) which may interfere with the reaction.

1. Blood Samples. Blood must be collected into ethylenediaminetetraacetic acid (EDTA) tubes and should be processed as soon as possible. Otherwise, keep blood samples refrigerated at 4 °C for no more than 2 weeks. For longer periods of preservation, addition of Guanidine-HCl buffer could be used (in a relation 1:1); this treatment will help preserve the DNA in blood samples for as long as 1 year at 4 °C, or 3 months at ambient temperature [6].

   Mononuclear cells purification. In order to attain a higher yield of DNA, buffy coat or mononuclear cells may be used as samples for DNA extraction. For diagnostic purpose we have purified mononuclear cells using the Hystopaque 1077 as follows:

   (a) Transfer carefully 3 ml of whole blood onto 3 ml of Hystopaque 1077 (it must be at ambient temperature before adding the blood). Centrifuge at 400 ×*g* for 30 min at ambient temperature.

   (b) Carefully aspirate and discard the upper layer (plasma), taking extra care not to disturb the opaque interface. Then transfer the mononuclear layer (the opaque interface) to a clean, labeled, 15 ml centrifuge tube.

   (c) Wash twice with 10 ml of isotonic phosphate buffered saline solution (PBS) and centrifuge at 1000 ×*g* for 10 min each time (at this point the process could be done at 4 °C).

   (d) Resuspend pelleted cells in 1 ml of PBS and transfer to a clean 1.5 ml centrifuge tube. Centrifuge at Vmax for 5 min.

   (e) Discard the supernatant and at this point you may store the sample at −20 or −70 °C until its use, or you could directly extract total DNA.

2. Bone Marrow Samples. Bone marrow samples should be stored in tubes containing 0.5 % EDTA as anticoagulant (must be properly collected by a physician) and processed as soon as possible. If stored, samples should be preserved at −20 °C until its use. Also, bone marrow aspirates can be collected on filter paper (Whatman no. 3 or FTA Cards), and then stored in separate plastic bags at −20 °C for further analysis. Bone marrow aspirates, as well as whole blood and spleen aspirates are commonly used samples, either for demonstration of parasites by smear evaluation or for molecular diagnosis. Bone marrow and spleen aspirates are obtained by invasive methodologies and may carry possible complications. Hence, to avoid invasive sampling methods, among other reasons, blood samples and molecular approaches have been more widely used lately for VL diagnosis [7].
3. Tissue Samples. Tissue aspirates from active lesions should be collected from the outer (infiltrated) border and base of the lesion, into syringes containing 0.5 ml of saline solution, 1000 U/ml penicillin, and 0.3 mg/ml streptomycin. Sample should be processed immediately or stored at −20 °C until further analysis [8]. Biopsy specimens are obtained from the active edge of the lesion with a sterile 4-mm diameter biopsy puncher, while for lesion scraping simple sterile lancets should be used. Samples should be stored in absolute ethanol at −20 °C until further use. For field work sampling, we recommend that tissue material is spotted onto FTA Classic Cards (Whatman Inc., Newton, MA, USA) [5], and samples be stored separately in plastic bags at ambient temperature for 18 months, or with silica gel at 4 °C for longer time.

When selecting whole blood (WB) versus buffy coat (BC) for DNA extraction, one should balance advantages and disadvantages between both methods. The use of WB is easier particularly in field studies and it has been reported to provide good sensitivity [9]. On the other hand, as *Leishmania* are obligate intracellular parasites (in the vertebrate host), they are expected to be more concentrated in the BC, thus this sample should yield a better sensitivity. Lachaud and coworkers in 2001 [10] performed a study comparing both samples for PCR diagnosis of VL, finding that BC develops a tenfold increase in sensitivity over that of WB. Therefore, as indicated by this study and our experience, we highly recommend using buffy coat or mononuclear purification strategies for Leishmaniasis diagnosis.

Because most areas of the world where Leishmaniasis is endemic are low-income rural areas with limited resources for adequate collection, storage, and transport of specimens, we have employed and highly recommend the use of FTA Cards, which are a filter paper that readily lyses the spotted material and fixes nucleic acids, making it a fast, simple, and reliable method for direct sampling [4, 5]. A similar sample collection technique has been described using filter paper (Whatman No. 3), with noticeably differences in the DNA extraction process and in diagnostic sensitivity [11]. The filter paper sampling technique has been widely used for Leishmaniasis diagnosis, recognized as a versatile and sensitive methodology, useful for field studies and to archive samples for very long periods of time and it is compatible with the use of many biological materials (blood, bone marrow, lymph aspirates and tissue aspirates or scraping as samples).

3.2 Preparation of Template for PCR

As for all infectious diseases, the limitation for diagnosis is the amount of the infecting organism in the host sample. In this specific case, it is the number of *Leishmania* parasites in the sample taken from the mammal host. The sensitivity of the diagnostic method will depend on the enrichment of the parasite in the sample; therefore it is important to select a proper sample and use an effective DNA extraction method.

For a better yield of DNA from blood samples we usually recommend organic extraction with phenol–chloroform–isoamyl alcohol (25:24:1), using either whole blood or buffy coat. However, depending on the number of samples to be processed, a blood DNA extraction kit may be used. The same recommendation stands for tissue DNA extraction, only that it undergoes a previous treatment with Proteinase K. Although the organic DNA extraction method in our experience gives a high yield of DNA, it does show some limitations as it uses hazardous organic solvents, is relatively time-consuming, and is almost impossible or prone to automate, being unsuitable for high-throughput applications [12]. Moreover, when using this method you should be aware of possible residual phenol or chloroform in the DNA, which may interfere with PCR reaction.

Leishmaniasis is a poverty-related disease, distributed mainly in rural areas of endemic countries. Thus, for epidemiologic studies, one must make sure to use a sampling method compatible with the conditions found in these areas, with basically no availability of refrigeration systems. In this regard, for molecular diagnosis of Leishmaniasis we recommend employing FTA Classic Cards which can be used directly for PCR and no previous DNA extraction process is needed.

1. From samples in filter paper (Adapted from Ref. 11).
   (a) Disks of filter papers with spotted biological material are punched out with a paper punch. After each sample is obtained, a clean sheet of paper sprayed with 90 % alcohol is punched ten times in order to prevent DNA contamination from one sample to the next.
   (b) Two disks (approximately 15 μl of aspirate) are placed in 250 μl of lysis buffer (50 mM NaCl, 50 mM EDTA pH 7.4, 1 % v/v-Triton X-100, and 200 μg of proteinase K per ml) and incubated for 3 h or overnight at 60 °C.
   (c) Samples are subjected to phenol-chloroform extraction, ethanol and Na-Acetate precipitation [13].

(d) DNA pellets are dried using a speed vacuum dryer for 5 min and redissolved in 50 μl of TE buffer (10 mM Tris 1 mM EDTA pH 7.5) or milli-Q water, DNA should be quantified in a spectrophotometer, and the quality of DNA measured at 260/280 nm. DNA samples are kept at −20 °C until its use.

2. From samples in FTA Classic Cards.

   (a) The biological material must be added onto the paper and let it dry completely. Then a 2 mm diameter disk is punched out from the filter paper for further processing.

   (b) Place the disk in a PCR tube and wash three times with the FTA Purification Reagent. Discard the used reagent after each wash.

   (c) Wash twice with TE buffer (10 mM Tris, 0.1 mM EDTA, pH 8.0) and discard the used buffer after each wash.

   (d) The disk is air dried in a PCR tube and subjected directly to PCR amplification.

Filter paper sampling method is a very useful tool for field work and easy to handle especially when working with many samples. When filter paper Whatman No. 3 is used, DNA extraction protocol is laborious, requiring the use of organic solvents, and at some point samples should be stored at 4 °C in order to attain better maintenance. In contrast, the use of FTA Cards provides a rapid and safe method without using any organic solvent or specialized equipment. When the samples are spotted onto an FTA card, the cells are lysed, pathogens are inactivated, and the nucleic acids are instantly fixed on the matrix of the card, resulting in protection from nuclease, oxidative and UV damage, and prevention of growth of bacteria and other microorganisms. This card is also suitable for long-term storage and the transportation is at room temperature [4].

3.3 Detection of Leishmania DNA by PCR

Diagnosis of Leishmaniasis commonly relies on culture methods, direct observation of parasites, or immunological techniques that detect specific antibodies in the sample. However, these techniques do not give any subtyping information of the infecting parasite. Therefore, molecular techniques like DNA-based analysis have been replacing the traditional methods, as they tend to be more specific and stable; and direct detection of molecular markers in host samples is being preferred. Since epidemiological studies normally involve large sample sets, methods that are cost-effective and allow high-throughput analyses are desirable [14].

There are many *Leishmania* species, and the different varieties of diseases are caused by specific species of the parasite. Hence, the ability to distinguish between *Leishmania* species is crucial for the correct diagnosis and prognosis of the disease, as well as for making proper decisions regarding treatment and control measures [14]. This is especially useful for epidemiological studies in areas with various co-existing *Leishmania* species, and those non-endemic areas were the *Leishmania* parasites have been imported due to the increasing international travel and population migration [15–17]. Hence, to differentiate *Leishmania* parasites at a species and strain levels, reliable, reproducible, and user-friendly tools are required.

Amplification of the parasites' DNA by the polymerase chain reaction (PCR) has evolved as one of the most specific and sensitive substitute methods for parasite detection in diagnosis of many infectious diseases. In the past two decades, many PCR applications for detecting *Leishmania spp.* have been reported, being the target regions usually multi-copy genes or polymorphic regions, in order to increase the probabilities of amplifying the parasites' DNA among a great amount of the host's DNA in the sample [18]. Some of the target regions described so far are the following: (1) small subunit ribosomal RNA gene (SSU-rRNA) [19], (2) kinetoplast DNA [20], (3) internal transcribed spacers (ITS) in the ribosomal operons [21], (4) heat shock protein 70 (hsp70) gene [22], (5) glucose-6-phosphate dehydrogenase gene [23], and (6) cytochrome *b* gene [24], among some others; each with different specificities and approaches in Leishmania diagnosis. Despite there are many PCR assays available, none has become a reference tool in Leishmanias diagnostics; therefore a method that ensures direct diagnosis and identification of pathogenic *Leishmania* species is still required for appropriate therapy programs to be developed [25, 26].

Since the advent of PCR, numerous molecular tools have been described to distinguish species and strains of *Leishmania* parasites. Here we show a review of the most common PCR-related protocols that are used to diagnose Leishmaniasis, identifying the species (Table 1). Depending on your needs and the equipment you have, you may choose the most suitable of these protocols. However, in respect of methods that can be used with samples taken in field work, we have successfully used and recommend the PCR protocol targeting the cytochrome *b* gene, collecting the samples into FTA Classic Cards for a better storage and handling [24, 27].

1. PCR amplification of the cytochrome *b* gene for *Leishmania spp.* species identification.

The cytochrome *b* (Cyt *b*) gene has been considered one of the most useful genes widely used for phylogenetic studies and identification of animals and plants. It is contained in the mitochondrial genome of a wide variety of living forms and encodes the central catalytic subunit of an enzyme present in the respiratory chain of mitochondria [35, 36].

In *L. tarentolae*, the Cyt *b* gene consists of two regions, the edited region (the 5′ region of 23 bp) that undergoes RNA editing, and the non-edited region (the 3′ region of 1056 bp). RNA editing process has been described in mitochondrial genes of kinetoplastid Protozoa [37]. In the RNA editing process, uridylate (U) residues are inserted or deleted to repair a frame-shift present

Table 1
Specifications of different molecular protocols for species-specific diagnosis of Leishmaniasis

Reference	Leishmaniasis	Leishmania species	Target (gene)	Technique	Specificity	Sensitivity
Da Graça et al. 2012 [26]	CL	*L. braziliensis, L. lainsoni, L. naiffi, L. guyanensis, L. shawi*, and *L. amazonensis*	hsp70	PCR–RFLP	72.7 %	73.2 %
Castilho et al, 2008 [23]	CL & MCL	*Leishmania (L.), Leishmania (V.), L braziliensis, L* no *braziliensis*	G6PD	Real-time PCR	NR	NR
Da Silveira Neto et al., 2012 [28]	VL	*L. infantum chagasi*	18S rRNA	PCR	NR	NR
Neitzke-Abreu et al., 2013 [29]	CL	*Leishmania* (Viannia)	MinicirclekDNA	PCR	PCR-L[a]: 100 %; PCR-Bb: 65.6 %	PCR-L[a]: 82.8 %; PCR-B[b]: 92.9 %
Volpini et al. 2004 [30]	ACL	*L.* (V.) *braziliensis* and *L.* (L.) *amazonensis*	Conserved regions of the minicirclekDNA	PCR-RFLP	98.5 %	NR
Adams et al. 2010 [31]	CL & VL	*L. donovani*	18S rRNA	RT-LAMP	VL: RT-LAMP 98 %	VL: RT-LAMP 83 %. CL:98 %
Khan et al. 2012 [32]	VL	*L. donovani*	SSU-rRNA	Nested PCR and LAMP	100 %	PCR: 96 %. LAMP: 90.7 %.
Toz et al. 2013 [33]	VL	*L. donovani complex, L. tropica*, and *L. major*	ITS1	RTq-PCR	NR	94.11 %

El Tail et al. 2000 [11]	VL and PKDL	*L. donovani*	ITS	PCR-SSCP	100 %	100 %
De Almeida et al. 2011 [34]		*Leishmania spp*	rRNA-ITS	PCR and DNA sequencing	NR	NR
Kato H. et al. 2005 [27]		*Leishmania spp*	Cyt*b*	PCR and DNA sequencing	NR	NR

VL visceral leishmaniasis, *CL* cutaneous lesihmaniasis, *MCL* mucocutaneous lesihmaniasis, *ACL* American cutaneous leishmaniasis, *PKDL* post-kala-azar dermal leishmaniasis, *hsp* heat shock protein, *G6PD* glucose 6-phosphate dehydrogenase, *rRNA* ribosomal RNA, *kDNA* kinetoplast DNA, *SSU-rRNA* small-subunit rRNA, *ITS* Internal transcribed spacer, *Cytb* Cytochrome *b gene*, *PCR* polymerase chain reaction, *RFLP* restriction fragment length polymorphism, *LAMP* loop-mediated isothermal amplification, *SSCP* single-strand conformation polymorphism, *NR* not reported

[a]PCR using DNA extracted from tissue as a template

[b]PCR using DNA extracted from leucocytes (from whole blood) as a template

in the genomic sequence and to create the AUG codon for translation initiation. Therefore, at the 3′ ends, the edited regions of 22 bp have a deletion of one T residue, while those of 24 bp have an insertion of one T residue.

In a study performed by Luyo-Acero et al. [24], they identified sequence variations of the Cyt *b* gene from human-infecting *Leishmania* species/subspecies, which has been widely used to distinguish each one of them in further studies. This approach has allowed exploring the phylogenic relationship among these parasites and has been a suitable and commonly used tool for diagnosis of Leishmaniasis [5, 38].

Cyt b-PCR protocol (Adapted from Refs. 4, 5).

We usually use nested PCR for amplification of *Leishmania* DNA from patient specimens on FTA Card to get maximum sensitivity [4, 7]. The protocol is as follows:

1. For the identification of *Leishmania* species, an analysis of the cytochrome *b* (*cyt* b) gene is performed through a nested PCR amplification with a pair of specific primers, L.cyt-AS (5′·GCGGAGAGRARGAAAAGGC·3′) and L.cyt-AR (5′·C CACTCATAAATATACTATA·3′).
2. Ampdirect Plus reagent (Shimadzu Biotech, Tsukuba, Japan) is to be used to obtain maximum sensitivity of PCR amplification against the possible carry-over of tissue-derived inhibitors for enzymatic reaction in each sample.
3. The suggested thermal profile is as follows: 30 cycles of 95.0 °C for 1 min, 55.0 °C for 1 min, and 72.0 °C for 1 min.
4. One microliter of the PCR product is subjected to the nested PCR with a set of inner primers, L.cyt-S (5′·GGTGTAGGTT TTAGTYTAGG·3′) and L.cyt-R (5′·CTACAATAAACAAAT CATAATATRCAATT·3′).

But, same samples such as your FTA samples from infected animals, DNA from parasite culture and infected sand fly specimens, single PCR using L.cyt-S and L.cyt-R primers is enough to amplify the parasite DNA. In many cases of subgenus *Viannia* species infection, parasite number in the lesion is lower in the chronic phase. Therefore, we are using nested PCR for patient specimens. In addition, we are using Ampdirect as a PCR buffer for FTA card and crude DNA as a template. General PCR solution works for FTA cards in most cases, but Ampdirect allows PCR reaction in the presence of enzyme inhibitors and increases the sensitivity.

The advantage of molecular approaches based on PCR or other amplification techniques is that they combine high sensitivity for direct detection of the infecting parasites in various human, animal, and sand fly tissues, with species specificity approaches. In areas where many species are sympatric, different approaches might be necessary, which would increase the cost of diagnostics. Approaches based on initial amplification of genus-specific sequences followed by subsequent differentiation of *Leishmania* species by RFLP, hybridization with specific probes or sequencing of the amplified sequences have proven most useful. However, the results of PCR diagnosis should always be evaluated in conjunction with clinical diagnosis in order to provide a better and reliable result.

References

1. Vega-López F (2003) Diagnosis of cutaneous leishmaniasis. Curr Opin Infect Dis 16(2):97–101
2. Paniz Mondolfi AE, Stavropoulos C, Gelanew T, Loucas E, Perez Alvarez AM, Benaim G, Polsky B, Schoenian G, Sordillo EM (2011) Successful treatment of Old World cutaneous leishmaniasis caused by Leishmania infantum with posaconazole. Antimicrob Agents Chemother 55(4):1774–1776
3. Paniz Mondolfi AE, Duffey GB, Horton LE, Tirado M, Reyes Jaimes O, Perez-Alvarez A, Zerpa O (2013) Intermediate/borderline disseminated cutaneous leishmaniasis. Int J Dermatol 52(4):446–455
4. Kato H, Cáceres AG, Mimori T, Ishimaru Y, Sayed ASM, Fujita M et al (2010) Use of FTA cards for direct sampling of patients' lesions in the ecological study of cutaneous leishmaniasis. J Clin Microbiol 48:3661–3665
5. Kato H, Watanabe J, Mendoza Nieto I, Korenaga M, Hashiguchi Y (2011) *Leishmania* species identification using FTA card sampling directly from patients' cutaneous lesions in the state of Lara, Venezuela. Trans R Soc Trop Med Hyg 105(10):561–567
6. Avila H, Sigman L, Cohen L, Millikan R, Simpson L (1991) Polymerase chain reaction amplification of *Trypanosoma cruzi* kinetoplast minicircle DNA isolated from whole blood lysates: diagnosis of chronic Chagas' disease. Mol Biochem Parasitol 48:211–221
7. Sundar S, Rai M (2002) Laboratory diagnosis of visceral leishmaniasis. Clin Diagn Lab Immunol 9:951–958
8. Boggild AK, Miranda-Verastegui C, Espinosa D, Arevalo J, Adaui V, Tulliano G, Llanos-Cuentas A, Low DE (2007) Evaluation of a microculture method for isolation of Leishmania parasites from cutaneous lesions of patients in Peru. J Clin Microbiol 45:3680–3684
9. Mathis A, Deplazes P (1995) PCR and *in vitro* cultivation for detection of *Leishmania spp.* in diagnostic samples from humans and dogs. J Clin Microbiol 33:1145–1149
10. Lachaud L, Chabbert E, Dubessay P, Reynes J, Lamothe J, Bastien P (2001) Comparison of various sample preparation methods for PCR diagnosis of visceral leishmaniasis using peripheral blood. J Clin Microbiol 39:613–617
11. El Tai NO, Osman OF, el Fari M, Presber W, Schönian G (2000) Genetic heterogeneity of ribosomal internal transcribed spacer in clinical samples of *Leishmania donovani* spotted on filter paper as revealed by single-strand conformation polymorphisms and sequencing. Trans R Soc Trop Med Hyg 94:575–579
12. Noguera N, Tallano C, Bragos I, Milani A (2001) Modified salting-out method for DNA isolation from newborn cord blood nucleated cells. J Clin Lab Anal 14:280–283
13. Meredith SE, Zijlstra EE, Schoone GJ, Kroon CC, van Eys GJ, Schaeffer KU, el-Hassan AM, Lawyer PG (1993) Development and application of the polymerase chain reaction for the detection and identification of *Leishmania* parasites in clinical material. Arch Inst Pasteur Tunis 70:419–431
14. Schönian G, Kuhls K, Mauricio IL (2011) Molecular approaches for a better understanding of the epidemiology and population genetics of *Leishmania*. Parasitology 138:405–425
15. Harms G, Schönian G, Feldmeier H (2003) Leishmaniasis in Germany. Emerg Infect Dis 9(7):872–875
16. Schönian G, Nasereddin A, Dinse N, Schweynoch C, Schallig HD, Presber W, Jaffe CL (2003) PCR diagnosis and characterization of Leishmania in local and imported clinical samples. Diagn Microbiol Infect Dis 47(1): 349–358
17. Johnston V, Stockley JM, Dockrell D, Warrell D, Bailey R, Pasvol G, Klein J, Ustianowski A, Jones M, Beeching NJ, Brown M, Chapman AL, Sanderson F, Whitty CJ (2009) Fever in returned travelers presenting in the United Kingdom: recommendations for investigation and initial management. J Infect 59:1–18
18. Antinori S, Calattini S, Longhi E, Bestetti G, Piolini R, Magni C, Orlando G, Gramiccia M, Acquaviva V, Foschi A, Corvasce S, Colomba C, Titone L, Parravicini C, Cascio A, Corbellino M (2007) Clinical use of polymerase chain reaction performed on peripheral blood and bone marrow samples for the diagnosis and

monitoring of visceral leishmaniasis in HIV-infected and HIV-uninfected patients: a single-center, 8-year experience in Italy and review of the literature. Clin Infect Dis 44:1602–1610

19. van Eys GJJM, Schoone GJ, Kroon NCM, Ebeling SA (1992) Sequence analysis of small subunit ribosomal RNA genes and its use for detection and identification of *Leishmania* parasites. Mol Biochem Parasitol 51: 133–142
20. Maurya R, Singh RK, Kumar B, Salotra P, Rai M, Sundar S (2005) Evaluation of PCR for diagnosis of Indian kala-azar and assessment of cure. J Clin Microbiol 43:3038–3041
21. Cupolillo E, GrimaldiJúnior G, Momen H, Beverley SM (1995) Intergenic region typing (IRT): a rapid molecular approach to the characterization and evolution of *Leishmania*. Mol Biochem Parasitol 73:145–155
22. Montalvo AM, Fraga J, Monzote L, Montano I, De Doncker S, Dujardin JC, Van der Auwera G (2010) Heat-shock protein 70 PCR-RFLP: a universal simple tool for Leishmania species discrimination in the New and Old World. Parasitology 137:1159–1168
23. Castilho TM, Camargo LM, McMahon-Pratt D, Shaw JJ, Floeter-Winter LM (2008) A real-time polymerase chain reaction assay for the identification and quantification of American Leishmania species on the basis of glucose-6-phosphate dehydrogenase. Am J Trop Med Hyg 78:122–132
24. Luyo-Acero G, Uezato H, Oshiro M, Kariya K, Katakura K, Gomez EAL, Hashiguchi Y, Nonaka S (2004) Sequence variation of the cytochrome *b* gene of various human infecting members of the genus *Leishmania* and their phylogeny. Parasitology 128:483–491
25. Deborggraeve S, Boelaert M, Rijal S, De Doncker S, Dujardin JC, Herdewijn P, Büscher P (2008) Diagnostic accuracy of a new Leishmania PCR for clinical visceral leishmaniasis in Nepal and its role in diagnosis of disease. Trop Med Int Health 13:1378–1383
26. Graça GC, Volpini AC, Romero GA, Oliveira Neto MP, Hueb M, Porrozzi R, Boité MC, Cupolillo E (2012) Development and validation of PCR-based assays for diagnosis of American cutaneous leishmaniasis and identification of the parasite species. Mem Inst Oswaldo Cruz 107(5):664–674
27. Kato H, Uezato H, Katakura K, Calvopiña M, Marco JD, Barroso PA, Gomez EA, Mimori T, Korenaga M, Iwata H, Nonaka S, Hashiguchi Y (2005) Detection and identification of Leishmania species within naturally infected sand flies in the Andean areas of Ecuador by a polymerase chain reaction. Am J Trop Med Hyg 72:87–93
28. Da Silveira Neto OJ, Duarte SC, da Costa HX, Linhares GF (2012) Design of primer pairs for species-specific diagnosis of *Leishmania (Leishmania) infantum chagasi* using PCR. Rev Bras Parasitol Vet 21:304–307
29. Neitzke-Abreu HC, Venazzi MS, Bernal MV, Reinhold-Castro KR, Vagetti F, Mota CA, Silva NR, Aristides SM, Silveira TG, Lonardoni MV (2013) Detection of DNA from *Leishmania (Viannia)*: accuracy of polymerase chain reaction for the diagnosis of cutaneous leishmaniasis. PLoS One 8:e62473. doi:10.1371/journal.pone.0062473
30. Volpini AC, Passos VM, Oliveira GC, Romanha AJ (2004) PCR-RFLP to identify *Leishmania (Viannia) braziliensis* and *L. (Leishmania) amazonensis* causing American cutaneous leishmaniasis. Acta Trop 90:31–37
31. Adams ER, Schoone GJ, Ageed AF, Safi SE, Schallig HD (2010) Development of a reverse transcriptase loop-mediated isothermal amplification (LAMP) assay for the sensitive detection of *Leishmania* parasites in clinical samples. Am J Trop Med Hyg 82:591–596
32. Khan MG, Bhaskar KR, Salam MA, Akther T, Pluschke G, Mondal D (2012) Diagnostic accuracy of loop-mediated isothermal amplification (LAMP) for detection of *Leishmania* DNA in buffy coat from visceral leishmaniasis patients. Parasit Vectors 5:280
33. Toz SO, Culha G, Zeyrek FY, Ertabaklar H, Alkan MZ, Vardarlı AT, Gunduz C, Ozbel Y (2013) A real-time ITS1-PCR based method in the diagnosis and species identification of Leishmania parasite from human and dog clinical samples in Turkey. PLoS Negl Trop Dis 7:e2205. doi:10.1371/journal.pntd.0002205
34. De Almeida ME, Steurer FJ, Koru O, Herwaldt BL, Pieniazek NJ, da Silva AJ (2011) Identification of *Leishmania spp.* by molecular amplification and DNA sequencing analysis of a fragment of rRNA internal transcribed spacer 2. J Clin Microbiol 49:3143–3149
35. Irwin DM, Kocher TD, Wilson AC (1991) Evolution of the cytochrome *b* gene of mammals. J Mol Evol 32:128–144
36. Degli Esposti M, De Vries S, Crimi M, Ghelly A, Patarnello T, Meyer A (1993) Mitochondrial cytochrome *b*: evolution and structure of the protein. Biochim Biophys Acta 1143: 243–271
37. Benne R (1994) RNA editing in trypanosomes. Eur J Biochem 221:9–23
38. Asato Y, Oshiro M, Myint CK, Yamamoto Y-i, Kato H, Marco JD, Mimori T, Gomez EAL, Hashiguchi Y, Uezato H (2009) Phylogenic analysis of the genus *Leishmania* by cytochrome *b* gene sequencing. Exp Parasitol 121:352–361

Chapter 12

Detection of *Trypanosoma cruzi* by Polymerase Chain Reaction

María Elizabeth Márquez, Juan Luis Concepción, Eglys González-Marcano, and Alberto Paniz Mondolfi

Abstract

American Trypanosomiasis (Chagas disease) is an infectious disease caused by the hemoflagellate parasite *Trypanosoma cruzi* which is transmitted by reduviid bugs. *T. cruzi* infection occurs in a broad spectrum of reservoir animals throughout North, Central, and South America and usually evolves into an asymptomatic chronic clinical stage of the disease in which diagnosis is often challenging. This chapter describes the application of polymerase chain reaction (PCR) for the detection of *Trypanosoma cruzi* DNA including protocols for sample preparation, DNA extraction, and target amplification methods.

Key words *Trypanosoma cruzi*, Chagas, Diagnosis, PCR

1 Introduction

American Trypanosomiasis (Chagas disease) continues to be among the most neglected of all tropical diseases, affecting millions of people and posing a serious health and economic setback for most Latin American countries [1, 2]. Estimates from the World Health Organization and the Centers for Disease Control and Prevention indicate that between 8 and 11 million people are infected, and at least 100 million are at risk of infection, by the disease-causing parasite *Trypanosoma cruzi* [1–3]. Traditional diagnosis of Chagas disease relies on demonstration of trypomastigotes in blood, amastigotes in tissues, serological testing, or culture [4]. However, most of these methods have important limitations; for example: circulating Trypomastigotes are usually absent in chronic stages of the disease, cultures and xenodiagnosis are cumbersome, and because most of the available serological tests use epimastigote antigens, these methods usually exhibit a high degree of cross-reactivity [4]. In this context, different PCR

Rajyalakshmi Luthra et al. (eds.), *Clinical Applications of PCR*, Methods in Molecular Biology, vol. 1392,
DOI 10.1007/978-1-4939-3360-0_12, © Springer Science+Business Media New York 2016

strategies have emerged as an important method for identification of *Trypanosoma cruzi*. Herein we describe a series of protocols as a combined strategy for the diagnosis of Chagas disease.

2 Materials

2.1 Equipment

1. Centrifuge.
2. Microcentrifuge.
3. Vortex.
4. Water bath.
5. PCR machine (Eppendorf Mastercycler gradient).
6. Agarose gel electrophoresis equipment.
7. Pipets for PCR.
8. Filter tips.
9. Sterilized tubes for master mix preparation (0.5, 1.5 ml), and reaction tubes.

2.2 Reagents

1. Axy Prep Blood Genomic DNA Miniprep Kit (Axygen Biosciences).
2. Phenol–Chloroform–isoamyl alcohol (25:24:1).
3. PCR kit.
4. Primers.
5. Milli-Q water.

3 Methods

3.1 Sample Nature and Storage

3.1.1 Blood Collection

Blood should be collected into ethylenediaminetetraacetic acid (EDTA) tubes, and processed as soon as possible. If not processed immediately, samples must be kept at 4 °C for no longer than 1 week until its use.

For prolonged storing, guanidine buffer (Guanidine–HCl 6 M pH 8.0, 0.2 mM EDTA) should be added to blood samples in a relation 1 to 1 (blood:buffer), and then stored at 4 °C. It has been reported that with Guanidine–HCl, blood could also be kept at ambient temperature for up to 3 months, which may be useful in field sample collection [5]. We have verified that in Guanidine–HCl treated blood samples, *Trypanosoma cruzi* DNA remains non-degraded for months, and we have successfully amplified *Trypanosoma* DNA from samples stored at 4 °C for 1.5 years. However, it should be noted that Guanidine–HCl is a salt which could inhibit PCR amplification and therefore dilutions of extracted DNA may be required for successful amplification.

3.1.2 Mononuclear Cells Purification

In order to attain better yields of DNA from whole blood samples, obtaining the buffy coat before extraction or purifying mononuclear cells is recommended. Both of these methods will enrich the DNA sample, and therefore increase the probabilities of amplifying the parasites DNA. For this purpose, we have used the Hystopaque 1077 solution as follows:

1. Transfer carefully 3 ml of whole blood onto 3 ml of Hystopaque 1077 (at ambient temperature). Centrifuge at 400 × *g* for 30 min at ambient temperature.
2. Carefully aspirate and discard the upper layer, taking extra care not to disturb the opaque interface. Then transfer the mononuclear layer (the opaque interface) to a clean, labeled, 15 ml centrifuge tube.
3. Wash twice the cells, with 10 ml of isotonic phosphate buffered saline solution (PBS) and centrifuge at 1000 × *g* for 10 min each time (at this point the process could be done at 4 °C).
4. Resuspend cell pellet in 1 ml of PBS and transfer to a clean 1.5 ml centrifuge tube. Centrifuge at Vmax for 5 min.
5. At this point you may store the sample at −20 or −70 °C until it is ready to use, or total DNA could be directly extracted.

3.1.3 Tissue Samples

For experimental procedures and human postmortem cases, tissue samples can be used for Chagas disease diagnosis. For DNA tissue extraction, fresh samples can be used or otherwise stored in liquid nitrogen or at −80 °C until its use. Paraffin embedded tissue may be used as well for diagnosis using the proper extraction method. The suggested protocol for DNA extraction from tissue samples is as follows [6]:

1. Homogenize organ samples and immediately mix with five volumes of lysis buffer (10 mM Tris–HCl pH 7.6, 10 mM NaCl, 0.5 % SDS, and 300 μg of proteinase K), incubating at 50 °C for about 18 h.
2. A total of 100 μl of phenol–chloroform–isoamyl alcohol (25:24:1) is added and then mix by vortex for 1 min and centrifuge for 4 min at 12,000 × *g*.
3. Transfer the aqueous phase to a clean tube and precipitate the DNA with ethanol and sodium acetate (3 M pH 5.5). Mix by vortex and then centrifuge at 12,000 × *g* for 10 min.
4. The pellet is washed with 70 % ethanol and resuspended in Tris–EDTA solution at pH 8.0 (TE buffer) or milli-Q water and stored at −70 °C until its use.

Diagnostic sensitivity can be enhanced by maximizing the amount of target DNA in the aliquot used for DNA extraction. This can be obtained by mixing blood specimens with guanidine

HCl–EDTA solution which lyses the parasites and releases their genetic content, thus making it possible to detect as little as one parasite in 20 ml of blood [5, 7]. It has also been reported that sensitivity could be enhanced by using blood clot as a primary specimen [8]. In addition, an alternate method consists in concentrating the parasites in the buffy coat fraction prior to DNA extraction [9]; this has been reported to significantly increase the sensitivity in comparison to frozen EDTA–blood and guanidine HCl–EDTA treated blood [10].

In a study developed by Qvarnstrom et al. [11], the authors compared PCR results obtained using buffy coat as primary specimen versus fresh EDTA–blood, and found that analysis using buffy coat allowed earlier detection and increasing levels of circulating parasitic genome material in three cases: two reactivation cases and a post-transplant acute infection. Combining both the buffy coat and whole blood aliquot in parallel can also provide helpful information to ensure test validity and to troubleshoot suspicious false positive PCR results. False positive PCR results obtained can be immediately flagged as suspicious, when the whole blood fraction is positive while the buffy coat is negative, therefore being advantageous to use both sample extraction processes. In such scenario, problem with the quality of the blood specimen, the DNA extraction process, or the PCR accuracy should be assessed.

Although promising, assessment about the advantages and robustness of analyzing both whole blood and buffy coat samples needs to be validated through further large studies. In conclusion, we propose that in reference laboratories with the adequate infrastructure, the use of two or more PCR tests with different performance characteristics combined with the analysis of buffy coat and whole blood can strengthen the use of PCR for accurate diagnosis of Chagas disease.

3.2 DNA Extraction with Phenol–Chloroform–Isoamylic Alcohol

1. Mix 100 μl of blood (mixed with Guanidine–HCl) with 100 μl of phenol–chloroform–isoamylic alcohol (25:24:1) (phenol Tris–EDTA pH 8, USB Corporation, USA).
2. Centrifuge for 3 min at 13,000 rpm, and then add 150 μl of distilled water, vortex, and centrifuge for 3 min at 13,000 rpm.
3. Transfer carefully the aqueous phase and add 200 μl of chloroform, vortex, and centrifuge for 3 min at 13,000 rpm.
4. Transfer the aqueous phase to a clean tube and mix with 40 μg of glycogen (from rabbit liver, Sigma, USA).
5. Precipitate DNA and glycogen with 200 μl of isopropanol, vortex and incubate for 35 min at −20 °C, centrifuge at 13,000 rpm for 15 min.
6. Wash the pellet adding 500 μl of ethanol 70 %, do not vortex, and centrifuge again for 15 min at 13,000 rpm.

7. Discard the ethanol and allow the pellet to dry for 10 min at 37 °C. Resuspend the pellet in 50 μl of TE buffer or sterile milli-Q water. DNA should be stored at −20 °C until its use.

The organic extraction method is a conventional technique for DNA extraction. First, the cells are lysed using a detergent which disrupts the cellular structure to create a lysate, which is followed by cell debris removal through centrifugation. Subsequently proteins are denatured using a protease, which together with lipids, carbohydrates, and other cell debris is later removed through extraction of the aqueous phase with the organic mixture of phenol, chloroform, and isoamylic alcohol [12]. The organic solvents precipitate the proteins leaving behind the polynucleotides (DNA and RNA), and the protein precipitate is then separated by centrifugation. In the last step, purified DNA is usually recovered by precipitation using ethanol or isopropanol (a property of ethanol precipitation is that it precipitates only polymeric nucleic acids and leaves behind short chain and monomeric nucleic acid components including the ribonucleotides from RNase treatment). The precipitated DNA is commonly resuspended in TE buffer or double distilled water, for further use in PCR amplification process [13]. The limitations of this method are that it uses hazardous organic solvents and is relatively time-consuming. Moreover, residual phenol or chloroform may be present in the extracted DNA, which can limit its use in downstream application such as PCR amplification. The process also generates toxic waste that must be disposed with care and in accordance with hazardous waste guidelines. In addition, although it generates a good yield of DNA, this technique is almost impossible to automate, making it unsuitable for high-throughput applications [14].

3.3 DNA Extraction Using a Commercial Kit

DNA is extracted from blood using the Axy Prep Blood Genomic DNA Miniprep Kit (Axygen Biosciences). The protocol is described subsequently, but for more details and troubleshooting, please refer to the kit handbook.

1. Add 500 μl of Buffer AP1 to a 1.5 ml microfuge tube.
2. Add 250 μl of anti-coagulated whole blood and mix by vortexing at top speed for 10 s (*see* **Note 1**).
3. Add 100 μl of Buffer AP2 and mix by vortexing at top speed for 10 s. Centrifuge at 12,000 × *g* for 10 min at ambient temperature to pellet cellular debris.
4. Place a Miniprep column into a 2 ml Microcentrifuge tube. Pipette the clarified supernatant obtained from **step 4** into the Miniprep column. Centrifuge at 12,000 × *g* for 1 min.
5. Discard the filtrate from the 2 ml Microfuge tube. Place the Miniprep column back into the 2 ml microfuge tube. Add 700

μl of Buffer W2 to the Miniprep column and allow it to stand at room temperature for 2 min. Centrifuge at 12,000 × *g* for 1 min (*see* **Note 2**).

6. Discard the filtrate from the 2 ml Microfuge tube. Place the Miniprep column back into the 2 ml microfuge tube. Add 800 μl of Buffer W2 to the Miniprep column and centrifuge at 12,000 × *g* for 1 min.

   Optional Step: Discard the filtrate from the 2 ml Microfuge tube. Place the Miniprep column back into the 2 ml Microfuge tube. Add 500 μl of Buffer W2 to the Miniprep column and centrifuge at 12,000 × *g* for 1 min.

7. Discard the filtrate from the 2 ml Microfuge tube. Place the Miniprep column back into the 2 ml Microfuge tube. Centrifuge at 12,000 × *g* for 1 min.
8. Place the Miniprep column into a 1.5 ml Microfuge tube. Add 100 μl of Buffer TE or milli-Q water. Allow to stand at room temperature for 1 min. Centrifuge at 12,000 × *g* for 1 min to elute the genomic DNA (*see* **Note 3**).

The concentration and quality of DNA should be measured spectrophotometrically at 260/280 nm in triplicate.

3.4 Comparison of DNA Isolation Methods

In a study developed by Ramírez et al. in 2009 [15], the authors compared the two proposed DNA isolation methods based on the efficiency of PCR amplification using the kinetoplast DNA (kDNA) and satellite DNA (stDNA) genomic regions when testing 100 positive samples that were ELISA, IIF, and TESA-blot positive, as well as ten negative control samples. Results of PCR efficiency amplification showed that the phenol–chloroform extraction method was 17 % more sensitive than the AquaPure Genomic DNA blood/tissue kit from Bio-Rad for the stDNA PCR, and 13 % more sensitive for the kDNA PCR. These results indicate that the differences between the two extraction methods using both PCR detection targets were statistically significant, and all the negative controls revealed an absence of amplification by both PCR methods. A statistically significant difference was observed when evaluating the DNA concentration, as the final DNA concentration obtained by the phenol–chloroform method was much higher than that obtained with the commercial kit. Therefore, as for Ramírez et al. [15] study and our experience in Chagas disease diagnosis, the phenol–chloroform DNA extraction method is the ideal method for *T. cruzi* DNA detection in blood or tissue samples. However, in a study performed by Schijman et al. [16] they determined that commercial DNA extraction kits offer better specificity than solvent extraction protocols.

3.5 PCR for Detection of Kinetoplast DNA from *Trypanosoma cruzi*

The most widely used PCR assays used for diagnostic purposes target either the kDNA, also called the minicircle, or a nuclear mini-satellite region designated TCZ. Both of these targets are present in multiple copies in the parasite genome, which increases the sensitivity of detection [17].

*T. cruzi*mitochondrial genome represents about 20 % of the parasite's total DNA and is organized as a concatenated network where thousands of small circular molecules, the minicircles, are found tightly interlocked, representing 95 % of the parasite's kDNA content. These molecules have characteristics that make them ideal targets for PCR detection, considering that they are present in an elevated number of copies (about 10,000) and that each minicircle contains four regions of highly conserved DNA sequences found in all strains and representative isolates of different *T. cruzi* lineages [18, 19].

PCR strategies using kDNA as an amplification target employ oligonucleotides designed for the conserved minicircle regions. As first described, the primers 121, 122 [20] amplify a 330 bp fragment of the minicircle kDNA which is the minicircle variable region [5, 21]. This approach has proven to be highly specific as it allows the successful detection of different *T. cruzi* strains, and also sensitive because it does not recognize other kinetoplastids [17, 22–25]. Studies using PCR assays targeting multicopy kDNA minicircles reported 100 % sensitivity in chronic Chagas disease patients, highlighting that an excess of human DNA does not interfere with the selective parasite DNA amplification process [26]. The sensitivity of the kDNA-PCR protocol is about 5 fg of total *T. cruzi* DNA in the reaction tube, equivalent to one parasite in 10 ml of blood [7, 27].

3.5.1 kDNA-PCR Protocol

1. An amplification reaction in a total volume of 25 μl, containing 250 ng of sample DNA, 1× *Taq* polymerase buffer, 3 mM of $MgCl_2$, 250 μM dNTP mix, 1 U of Go *Taq* polymerase, 0.25 μM of each forward and reverse primers (121: 5′·AAATAATG TACGGGKGAGATGCATGA·3′ and 122: 5′·GGTTCGATT GGGGTTGGTGTAATATA·3′ respectively) and a quantity of Milli-Q water sufficient to complete the final volume.
2. Suggested thermal profile:

	94.0 °C 3 min
5 cycles of:	94.0 °C 1 min
	68.0 °C 1 min
	72.0 °C 1 min

35 cycles of:	94.0 °C 45 s 64.0 °C 45 s 72.0 °C 45 s
	72.0 °C 10 min

3. An amplification reaction without DNA template is included in each PCR round as a negative control to exclude false positives resulting from contamination of reagents. A *T. cruzi* positive sample from a patient and genomic DNA from *T. cruzi* strain Y are included as positive controls.
4. The presence of an amplification product (kDNA 330 bp) is analyzed in a 2 % agarose gel, detected with ethidium bromide (Fig. 1).

3.6 PCR for Detection of Satellite DNA from *Trypanosoma cruzi*

T. cruzi satellite DNA (Sat-DNA) is present in 120,000 copies in the parasite genome, is a 195-bp repeated sequence, and it represents 10 % of the parasites' total DNA, which makes it a highly sensitive target [28, 29]. To establish the sensitivity of sat-DNA

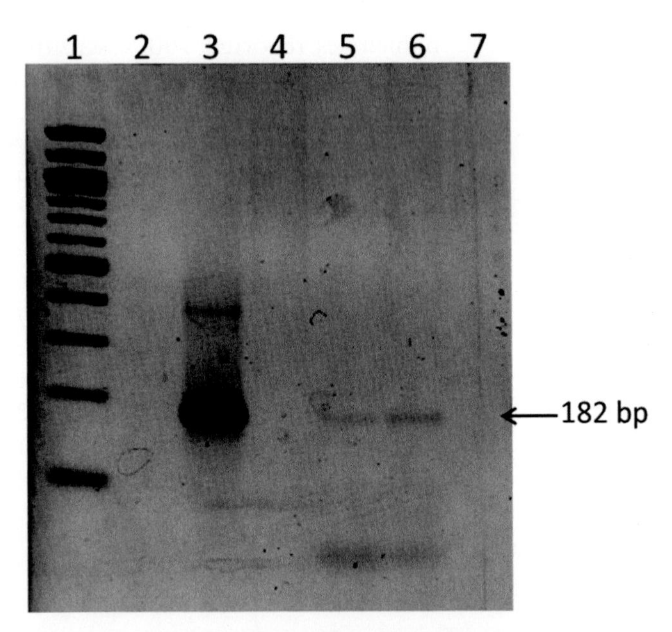

Fig. 1 Polymerase chain reaction (PCR) amplification for the kinetoplast DNA gene (kDNA) of *Trypanosoma cruzi*. Amplification bands are visualized on a 2 % agarose gel stained with ethidium bromide. *Lanes*: (*1*) 100 bp DNA ladder, each lane corresponds to 100 base pairs, (*2*) Negative control without DNA, (*3*) Positive control with *T. cruzi* genomic DNA, (*4*) DNA from an infected dog post-treatment, (*5*) DNA from an infected dog, (*6*) DNA from an infected dog, (*7*) DNA from a non-infected dog. All DNA samples from dogs were extracted from whole blood

amplification primed by TCZ1 and TCZ2 primers, genomic DNA from various sources and templates were used (Y, Tulahuén, Corpus Christi, and Sylvio X-10/4 isolates) and obtained from patients in widely separated geographical areas (Brazil, Chile, and Texas) yielding a 188 bp amplification band; thus suggesting that a fragment size is universally present in these parasites. Moreover, TCZ1 and TCZ2 are highly specific, because they do not amplify DNA of closely related species, as *Leishmania spp.* (*L. mexicana*, *L. major*, *L. braziliensis*, *L. donovani*) which overlap with *T. cruzi* in some endemic areas, as well as the African trypanosomes (*T. brucei brucei*, *T. brucei rhodesiense*, *T. brucei gambiense*, and *T. congolense*) which are also phylogenetically related to *T. cruzi* [30]. Additionally, mammalian hosts of *T. cruzi* such as mice or humans do not have DNA sequences that are amplified to any significant degree with these primers [29]. Sat-DNA PCR tests showed high specificity and sensitivity values of 0.05–0.5 parasites/ml, whereas specific kDNA tests detected 5×10^{-3} parasites/ml [16]. Studies conducted in infected monkeys and in patients using serum samples to amplify Sat-DNA revealed a high sensitivity and specificity in the detection of DNA from *T. cruzi* at any stage of the disease [31].

3.6.1 Sat-DNA PCR Protocol

1. An amplification reaction in a total volume of 25 μl, containing 250 ng of sample DNA, 1× *Taq* polymerase buffer, 3 mM of $MgCl_2$, 250 μM dNTP mix, 1 U of *Taq* polymerase, 0.5 μM of each forward and reverse primers (TCZ1: 5′-GCTCTTGCCCACAMGGGTGC-3′ TCZ2: 5′-CCAAGCAGCGGATAGTTCAGG-3′ respectively), and a quantity of Milli-Q water sufficient to complete the final volume.
2. Suggested thermal profile:

	98.0 °C 10 min
40 cycles of:	98.0 °C 45 s
	55.5 °C 1 min
	72.0 °C 1 min
	72.0 °C 15 min

3. An amplification reaction without DNA template is included in each PCR round as a negative control to exclude false positives resulting from contamination of reagents. A *T. cruzi* positive sample from a patient and genomic DNA from *T. cruzi* strain Y are included as positive controls.
4. The presence of an amplification product (Sat-DNA 182 bp) is analyzed in a 2 % agarose gel, detected with ethidium bromide (Fig. 2).

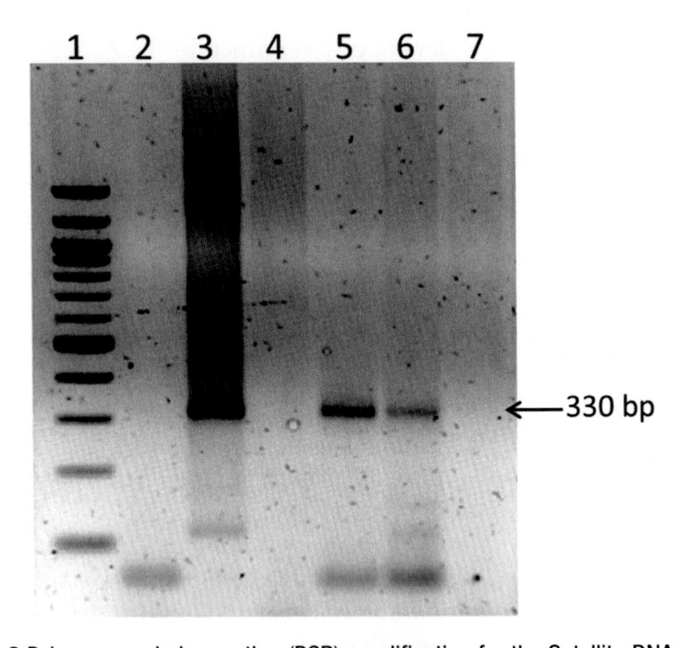

Fig. 2 Polymerase chain reaction (PCR) amplification for the Satellite DNA gene (Sat-DNA) of *Trypanosoma cruzi*. Amplification bands are visualized on a 2 % agarose gel stained with ethidium bromide. *Lanes*: (*1*) 100 bp DNA ladder, each lane corresponds to 100 base pairs, (*2*) Negative control without DNA, (*3*) Positive control with *T. cruzi* genomic DNA, (*4*) DNA from an infected dog post-treatment, (*5*) DNA from an infected dog, (*6*) DNA from an infected dog, (*7*) DNA from a non-infected dog. All DNA samples from dogs were extracted from whole blood

3.7 PCR for Detection of Tc24 Gene from *Trypanosoma cruzi*

In a study by Guevara et al. [32] the authors experimentally infected mice with *T. cruzi* trypomastigotes and followed up on blood samples periodically, as the mice were sacrificed at 4, 15, 20, 28, and 120 days post-infection (PI). DNA was isolated from blood and a variety of different tissues (heart, liver, kidney, duodenum, ileum, spleen, colon, and skeletal muscle). Though parasites were detected by PCR (Tc24-based PCR) in all tissues at different times, the highest number of parasitism and the better resolved PCR bands were most clearly observed at day 28th PI, with skeletal muscle and heart showing the highest intensity of PCR amplified products, thus suggesting the presence of large numbers of parasites. These results were confirmed by histological analysis for the presence of amastigotes nests. Although parasites were present in other organs, they were mostly localized within macrophages and no extensive tissue damage was observed in comparison with heart and skeletal muscle. Based on the abovementioned findings and the results obtained by us using this same protocol, we have observed that the Tc24-based PCR assay is more sensitive than the *T. cruzi* kDNA amplification based method, making it more suitable for detection on tissue samples. In addition, studies by Taibi et al. [33] propose that the Tc24-based PCR assay is a suitable method for making a differential diagnosis between *T. cruzi* and *T. rangeli*.

3.7.1 Tc-24 PCR Protocol

Detection of the DNA sequence encoding the Tc-24 protein of *T. cruzi* was used for the identification of parasites in blood and tissue samples. Amplification of Tc-24 sequence was performed using one set of primers for a first PCR as reported by Ouaissi et al. [34].

1. The first PCR mix was prepared to a final volume of 50 μl as follows: 5 μl of 10× Taq polymerase reaction buffer, 2 μl of 2.5 mM dNTP's solution, 3 μl of 50 mM $MgCl_2$, 1 μl of DNA sample (100 ng), and 7 μl of each pair of 3 μM oligonucleotides (sense primer 5′ · GACGGCAAGAACGCCAAGGAC · 3′ and antisense primer 5′ · TCACGCGCTCTCCGGCACGTTG TC · 3′), 1 U of Taq DNA polymerase and water sufficient to complete the final volume.
2. To improve the specificity, a nested PCR was performed designing an internal primer sequence of Tc-24. This reaction was performed as follows: 2 μl of the first PCR amplification product and 7 μl of each pair of 3 μM oligonucleotides (sense 5′ · AAGAAGTTCGACAAGAACGA · 3′ and antisense 5′ · AAACTCGTCGAACGTCACGG · 3′) in a 50 μl reaction.
3. The suggested thermal profile:

	94.0 °C 5 min
35 cycles of:	94.0 °C 1 min 62.0 °C 1 min
	72.0 °C 2 min
25 cycles (nested PCR):	94.0 °C 1 min 50.0 °C 1 min
	72.0 °C 1 min
	72.0 °C 7 min

4. PCR product of 550 bp was detected in 2 % agarose gel electrophoresis with ethidium bromide (Fig. 3).

3.8 Utility of PCR Strategies for Diagnosis of Chagas Disease

The application of PCR techniques from blood or tissue samples has raised new diagnostic and prognostic possibilities. Research on the use of a PCR technique based on the 330 bp kDNA and the 195 bp nuclear DNA satellite (Sat-PCR) repetitive sequences allows early diagnosis of *T. cruzi* infection recurrence in heart transplantation (HTx) patients [35]. Indeed, PCR enabled the detection of parasite DNA about 41 days prior to plasma parasite detection by conventional methods. Long-term follow-up of ten consecutive *T. cruzi* infected cases showed kDNA-PCR positive findings around 60 days prior to clinical reactivation. Additionally, PCR has been a potentially useful tool for diagnosis of *T. cruzi* infection in newborn to infected mothers, which is essential for the rapid administration of anti-parasitic drugs. Virreira et al. [36]

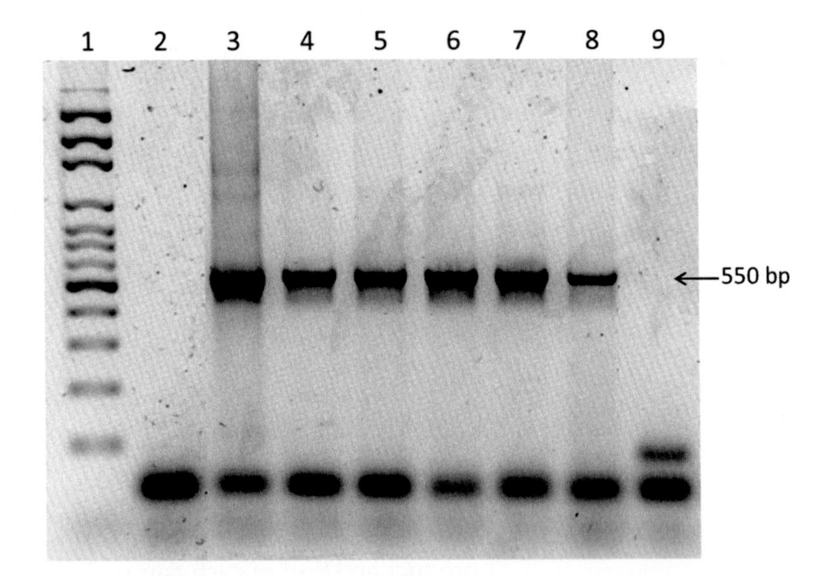

Fig. 3 Polymerase chain reaction (PCR) amplification for the detection of the DNA sequence encoding the Tc-24 protein from *Trypanosoma cruzi*. Amplification bands are visualized on a 2 % agarose gel stained with ethidium bromide. *Lanes*: (*1*) 100 bp DNA ladder, each lane corresponds to 100 base pairs, (*2*) Negative control without DNA, (*3*) Positive control with *T. cruzi* genomic DNA, (*4–8*) DNA extracted from heart tissue of mice infected with *T. cruzi*, (*9*) DNA extracted from heart tissue of a mouse infected with *T. cruzi*

determined that the Sat-PCR protocol detects a single parasite in 0.1 ml of blood, and such sensitivity should be sufficient to detect congenital infection corresponding to an acute parasitic phase, also showing a good specificity in newborns from infected and non-infected mothers. Therefore, this technique may be a useful alternative tool in the diagnosis of congenital Chagas disease.

Treatment research for Chagas disease has been considered a promising medical approach to patients in either the acute, indeterminate, or chronic phases. It is necessary to be able to monitor the efficacy of several therapeutic alternatives and to establish reliable criteria to determine the parasitological cure of patients. The high sensitivity and specificity of the PCR-based diagnosis of *T. cruzi* infection makes it a suitable tool for the follow-up of a chemotherapeutic treatment of chagasic patients. Therefore, the method could be used to assess parasite clearance from blood or cardiac tissue [37].

3.9 Loop-Mediated Isothermal Amplification (LAMP) Technique

Loop-mediated isothermal DNA amplification (LAMP) is a powerful innovative gene amplification technique emerging as a simple rapid diagnostic tool for early detection of several infectious diseases, including viral, bacterial, and parasitic diseases [38]. The mechanism of the LAMP reaction can be simplified in three steps: an initial step, a cycling amplification step, and an elongation step.

LAMP employs a DNA polymerase with strand-displacement activity (*Bst* DNA polymerase, obtained *Bacillus stearothermophilus*), along with two inner primers (FIP, BIP) and outer primers (F3,B3) which recognize six separate regions within a target DNA sequence. The LAMP assay amplifies specific sequences of DNA under isothermal conditions in the range of 65 °C, and as a result it allows the use of simple and no special reagent, no sophisticated equipment is required, it has high specificity because the amplification reaction occurs only when all six regions within a target DNA are correctly recognized by the primers, it amplifies a specific gene from a genome discriminating a single nucleotide difference, and shows high amplification efficiency (enables amplification in a shorter time) that is attributed to no time loss of thermal change because of its isothermal reaction [39]. It also produces tremendous amount of amplified products that makes a simple detection possible [40].

3.9.1 LAMP for Detection of rRNA Genes from ***Trypanosoma cruzi***

1. The LAMP primer sets were designed from 18S rRNA genes of *T. cruzi* [41]. The design and operation of the two outer primers (F3 and B3) is the same as that of the regular PCR primers. Each of the inner primers (FIP and BIP) contains two distinct sequences that correspond to the sense (FIPF2 and BIP-B2) and the antisense (FIP-F1c and BIP-B1c) sequences of the target DNA, and they form stem-loop structures at both ends of the minimum LAMP reaction unit. These stem-loop structures initiate self-primed DNA synthesis and serve as the starting material for subsequent LAMP cycling reaction [42].

3.9.2 18S rRNA Genes LAMP Protocol

For LAMP reactions, the material (reaction tubes) and reagents (Loopamp Fluorescent Detection Reagent) provided by the commercial kit are used.

1. LAMP reaction mixture in a total volume of 25 μl contained: 12.5 μl of the reaction buffer (40 mM Tris–HCl pH 8.8, 20 mM KCl, 16 mM $MgSO_4$, 20 mM $(NH_4)_2SO_4$, 0.2 % Tween 20, 1.6 M Betaine, 2.8 mM of each dNTP), between 200 and 400 ng of sample DNA extracted from blood or tissue, 8 U of *Bst* DNA polymerase, 6 μl of primer mix (FIP: 5′·GGTAAAAA ACCCGGCTTTCGCAACCGGCAGTAACACTCAGA·3′, BIP: 5′·CGATGGCCGTGTAGTGGACTGTTTCTCAGGC TCCCTCTCC·3′ at 40 pmol each; F3: 5′·GGACGTCCAGCGAATGAATG·3′ and B3: 5′·CCTCCGTAGAAGTGGTAGCT·3′) at 5 pmol each; Loop-F and Loop-B at 20 pmol each, 1 μl Fluorescent Detection Reagent and water sufficient to complete the final volume.
2. The mixture is incubated at 64 °C for 40 min and then at 85 °C for 5 min.
3. An amplification reaction without DNA template is included in each PCR round as a negative control to exclude false positives

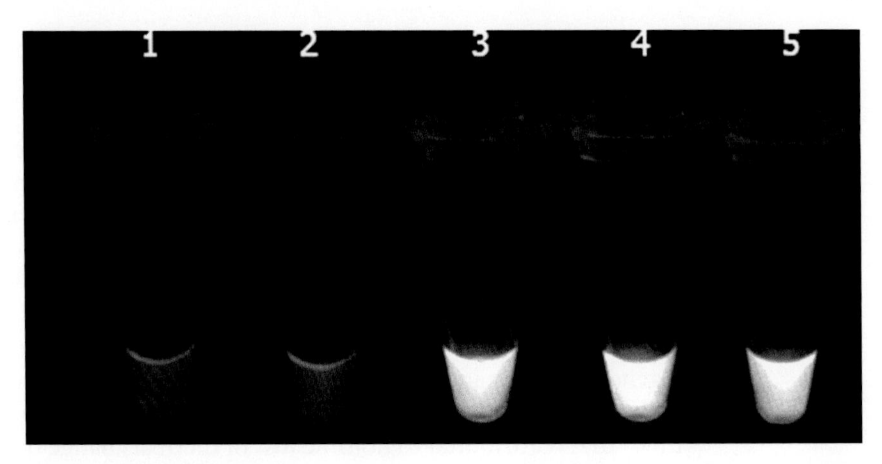

Fig. 4 Loop-mediated isothermal DNA amplification (LAMP) for the detection of the 18S rRNA genes from *Trypanosoma cruzi*. Amplification product is visualized directly on the reaction tubes using a molecular imager. *Tubes*: (*1*) Negative control without DNA, (*2*) 760 ng of DNA extracted from heart tissue of a non-infected rat, (*3*) 760 ng of DNA extracted from heart tissue of an infected rat, (*4*) 1 pg of genomic DNA from *T. cruzi*, (*5*) 140 ng of DNA extracted from blood of a patient with Chagas disease in acute phase

resulting from contamination of reagents. A *T. cruzi* positive sample from a patient and genomic DNA from *T. cruzi* are also included as positive controls.

4. Amplification product is observed in a Molecular Imager ChemiDoc XRS (BioRad) (Fig. 4).

In a study developed by Concepción et al. (not published) the LAMP technique was used to detect *T. cruzi* DNA in samples of rat heart tissue and human blood, using as a negative control DNA extracted from heart of a non-infected rat, and as a positive control DNA from cultured *T. cruzi* epimastigotes. As shown in Fig. 4, no signal was observed when no DNA was added to the mix and in the negative control, however, a strong signal was detected in the positive samples, confirming the high sensitivity of the LAMP method.

3.9.3 Advantages of the LAMP Technique

LAMP is an established nucleic acid amplification method that offers rapid, accurate, and cost-effective diagnosis of infectious diseases [43]. This molecular tool can be used for rapid field diagnosis in rural areas where there is no access to sophisticated equipment for detection. This technique eradicates the need for expensive thermocyclers used in conventional PCR; it may be a particularly useful method for infectious disease diagnosis in the non-developed countries. An additional advantage of LAMP over PCR-based methods is that DNA amplification can be monitored spectrophotometrically and/or with the naked eye without the use of special dyes [44].

LAMP assay allows the detection of parasitic infections. Thekisoe et al. in 2010 [45] developed two loop-mediated isothermal amplification (LAMP) assays for specific detection of *Trypanosoma cruzi* and *Trypanosoma rangeli* based on the 18S ribosomal RNA (rRNA) and the small nucleolar RNA (snoRNA) genes, respectively. The *T. cruzi* 18S LAMP assay specifically amplifies *T. cruzi* DNA without amplifying negative control with *T. rangeli*, vector insects, and human host DNA. The detection limit of the assay was 100 fg of serially diluted *T. cruzi* DNA. In this study, six LAMP primers were used for amplification of each target trypanosome DNA. In this sense eight distinct regions were recognized on the target gene, thereby ensuring specificity, high sensitivity, and rapid reaction whereby amplification is achieved within 60 min. This result is a very important approach as the nonpathogenic *Trypanosoma rangeli* shares the same geographical location and same insect vectors with *T. cruzi* parasites, hence being necessary an accurate differential diagnosis. This study brings LAMP to the forefront as an alternative for molecular diagnosis and confirmation of differential *T. cruzi* and *T. rangeli* infections in vectors, clinical samples, transfusion blood samples, and organs for transplantation.

4 Notes

1. Vortexing is required for complete release of the genomic DNA. Although vortexing will result in limited shearing of the genomic DNA, it will have no effect upon the performance of the genomic DNA in applications which require high molecular DNA.
2. If any liquid remains in the Miniprep column after centrifugation, extend the centrifuge time or increase the *g*-force.
3. Pre-warming Buffer TE or milli-Q water at 65 °C will generally improve elution efficiently.

References

1. Benaim G, Paniz Mondolfi AE (2012) The emerging role of amiodarone and dronedarone in Chagas disease. Nat Rev Cardiol 9(10):605–609
2. Hotez PJ, Dumonteil E, Woc-Colburn L, Serpa JA, Bezek S, Edwards MS, Hallmark CJ, Musselwhite LW, Flink BJ, Bottazzi ME (2012) Chagas disease: “The new HIV/AIDS of the Americas”. PLoS Negl Trop Dis 6(5):e1498
3. Centers for Disease Control and Prevention (2011) Chagas disease in the Americas – 2011 (online)
4. Versalovic J, Carroll KC, Funke G, Jorgensen JH, Landry ML, Warnock DW (eds) (2011) MCM: manual of clinical microbiology, 10th edn. ASM Press, Washington, DC. ISBN 978-1-55581-463-2
5. Avila H, Sigman L, Cohen L, Millikan R, Simpson L (1991) Polymerase chain reaction amplification of Trypanosoma cruzi kinetoplast minicircle DNA isolated from whole blood lysates: diagnosis of chronic Chagas’ disease. Mol Biochem Parasitol 48:211–221
6. Vera J, Magallón E, Grijalva G, Ramos C, Armendáriz B (2003) Molecular diagnosis of

Chagas' disease and use of an animal model to study parasite tropism. Parasitol Res 89(6):480–486

7. Britto C, Cardoso M, Wincker P, Morel C (1993) A simple protocol for the physical cleavage of Trypanosoma cruzi kinetoplast DNA present in blood samples and its use in polymerase chain reaction (PCR)-based diagnosis of chromic Chagas' disease. Mem Inst Oswaldo Cruz 88:171–172
8. Fitzwater S, Calderon M, Lafuente C, Galdos G, Ferrufino L, Verastegui S, Gilman R, Bern C (2008) Polymerase chain reaction for chronic Trypanosoma cruzi infection yields higher sensitivity in blood clot than buffy coat or whole blood specimens. Am J Trop Med Hyg 79:768–770
9. Feilij H, Muller L, Gonzalez C (1983) Direct micromethod for diagnosis of acute and congenital Chagas' disease. J Clin Microbiol 18:327–330
10. Fernandes C, Tiecher F, Balbinot M, Liarte D, Scholl D, Steindel M, Romanha A (2009) Efficacy of benznidazol treatment for asymptomatic chagasic patients from state of Rio Grande do Sul evaluated during a three years follow-up. Mem Inst Oswaldo Cruz 104(1):27–32
11. Qvarnstrom Y, Schijman A, Veron V, Aznar C, Steurer F, da Silva A (2012) Sensitive and specific detection of Trypanosoma cruzi DNA in clinical specimens using multi-target real-time PCR approach. PLoS Negl Trop Dis 6(7):e1689
12. Guevara A, Eras J, Recalde M, Vinueza L, Cooper P, Ouaissi A, Guderian R (1997) Severe digestive pathology associated with chronic Chagas' disease in Ecuador: report of two cases. Rev Soc Bras Med Trop 30(5):389–392
13. Buckingham L, Flaws M (2007) Molecular diagnostics: fundamentals, methods and clinical applications. F.A Davis, Philadelphia, PA
14. Noguera N, Tallano C, Bragos I, Milani A (2001) Modified salting-out method for DNA isolation from newborn cord blood nucleated cells. J Clin Lab Anal 14:280–283
15. Ramírez J, Guhl F, Umezawa E, Morillo C, Rosas F, Marin J, Restrepo S (2009) Evaluation of adult chronic Chagas' heart disease diagnosis by molecular and serological methods. J Clin Microbiol 47(12):3945–3951
16. Schijman A, Bisio M, Orellana L, Sued M, Duffy T, Mejia A, Cura C, Auter F, Veron V, Qvarnstrom Y, Deborggraeve S, Hijar G, Zulantay I, Horacio R, Velazquez E, Tellez T, Sanchez Z, Galvao L, Nolder D, Rumi M, Levi J, Ramirez J, Zorrilla P, Flores M, Jercic M, Crisante G, Añez N, De Castro A, Gonzalez C, Acosta K, Yachelini P, Torrico F, Robello C, Diosque P, Triana O, Aznar C, Russomando G, Buscher P, Assa A, Guhl F, Sosa S, DaSilva S, Britto C, Luquetti A, Ladzins J (2011) International study to evaluate PCR methods for detection of Trypanosoma cruzi DNA in blood samples from Chagas disease patients. PLoS Negl Trop Dis 5(1):e931
17. Sturm N, Degrave W, Morel C, Simpson L (1989) Sensitive detection and schizodeme classification of Trypanosoma cruzi cells by amplification of kinetoplast minicircle DNA sequences: use in diagnosis of Chagas' disease. Mol Biochem Parasitol 33:205–214
18. Degrave W, Fragoso S, Britto C, van Heuverswyn H, Kidane G, Cardoso M, Mueller R, Simpson L, Morel C (1988) Peculiar sequence organization of kinetoplast DNA minicircles from Trypanosoma cruzi. Mol Biochem Parasitol 27(1):63–70
19. Simpson L (1997) The genomic organization of guide RNA genes in kinetoplastid protozoa: several conundrums and their solutions. Mol Biochem Parasitol 86(2):133–141
20. Wincker P, Bosseno M, Britto C, Yaksic N, Cardoso M, Morel C, Breniere S (1994) High correlation between Chagas' disease serology and PCR based detection of Trypanosoma cruzi kinetoplast DNA in Bolivian children living in an endemic area. FEMS Microbiol Lett 124:419–423
21. Gomes M, Galvao L, Macedo A, Pena S, Chiari E (1999) Chagas' disease diagnosis: comparative analysis of parasitologic, molecular, and serological methods. Am J Trop Med Hyg 60:205–210
22. Vallejo G, Guhl F, Chiari E, Macedo A (1999) Species specific detection of Trypanosoma cruzi and Trypanosoma rangeli in vector and mammalian hosts by polymerase chain reaction amplification of kinetoplast minicircle DNA. Acta Trop 72:203–212
23. Guhl R, Vallejo G (2003) Trypanosoma (Herpetosoma) rangeli Tejera, 1920- an updated review. Mem Inst Oswaldo Cruz 98(4):435–442
24. Elías C, Vargas N, Zingales B, Schenkman S (2003) Organization of satellite DNA in the genome of Trypanosoma cruzi. Mol Biochem Parasitol 129:1–9
25. Rodríguez I, Marína C, Pérez G, Gutiérrez R, Sánchez M (2008) Identification of trypanosoma strains isolated in Central and South America by endonucleases cleavage and duplex PCR of kinetoplast-DNA. Open Parasitol J 2:35–39

26. Avila H, Pereira J, Thiemann O, De Paiva E, Degrave W, Morel C, Simpson L (1993) Detection of Trypanosoma cruzi in blood specimens of chronic patients by polymerase chain reaction amplification kinetoplast minicircle DNA: comparison with serology and xenodiagnosis. J Clin Microbiol 31:2421–2426
27. Diez M, Favaloro L, Bertolotti A, Burgos J, Vigliano C, Lastra M, Arnedo A, Nagel C, Schijman A, Favaloro R (2007) Usefulness of PCR strategies for early diagnosis of Chagas' Disease reactivation and treatment follow-up in heart transplantation. Am J Transplant 7:1633–1640
28. Kooy R, Ashall F, Van der Ploeg M, Overdulve J (1989) On the DNA content of *Trypanosoma cruzi*. Mol Biochem Parasitol 36:73–76
29. Gonzalez A, Prediger E, Huecas M, Nogueira N, Lizardi P (1984) Minichromosomal repetitive DNA in Trypanosoma cruzi: its use in a high sensitivity parasite detection assay. Proc Natl Acad Sci U S A 81:3356–3360
30. Moser D, Kirchhoff L, Donelson J (1989) Detection of Trypanosoma cruzi by DNA amplification using the polymerase chain reaction. J Clin Microbiol 27:1477–1482
31. Russomando G, Figueredo A, Almiron M, Sakamoto M, Morita K (1992) Polymerase chain reaction based detection of Trypanosoma DNA in serum. J Clin Microbiol 30(11):2864–2868
32. Guevara E, Taibi A, Bilaut M, Ouaissi A (1996) PCR-based detection of *Trypanosoma cruzi* useful for specific diagnosis of human Chagas' disease. J Clin Microbiol 34:485–486
33. Taibi A, Guevara E, Schoneck R, Yahiaoui B, Ouiaissi A (1995) Improved specificity of Trypanosoma cruzi identification by polymerase chain reaction using an oligonucleotide derived from the amino-terminal sequence of a Tc24 protein. Parasitology 111:581–590
34. Ouaissi A, Aguirre T, Plumas-Marty B, Piras M, Schöneck R, Gras-Masse H, Taibi A, Loyens M, Tartar A, Capron A et al (1992) Cloning and sequencing of a 24-kDa Trypanosoma cruzi specific antigen released in association with membrane vesicles and defined by a monoclonal antibody. Biol Cell 75(1):11–17
35. Schijman AG, Vigliano C, Burgos J, Favaloro R, Perrone S, Laguens R, Levin MJ (2000) Early diagnosis of recurrence of Trypanosoma cruzi infection by polymerase chain reaction after heart transplantation of a chronic Chagas' heart disease patient. J Heart Lung Transplant 19(11):1114–1117
36. Virreira M, Torrico F, Truyens C, Alonso C, Solano M, Carlier Y, Svoboda M (2003) Comparison of polymerase chain reaction methods for reliable and easy detection of congenital *Trypanosoma cruzi* infection. Am J Trop Med Hyg 68(5):574–582
37. Wincker P, Britto C, Pereira JB, Cardoso MA, Oeleman O, Morel CM (1994) Use of a simplified polymerase chain reaction procedure to detect Trypanosoma cruzi in blood samples from chronic chagasic patients in a rural endemic area. Am J Trop Med Hyg 51:771–777
38. Iseki H, Kawai S, Takahashi N, Hirai M, Tanabe K, Yokoyama N, Igarashi I (2010) Evaluation of a loop-mediated isothermal amplification method as a tool for diagnosis of infection by the zoonotic simian malaria parasite Plasmodium knowlesi. J Clin Microbiol 48(7):2509–2514
39. Parida M, Sannarangaiah S, Dash P, Rao P, Morita K (2008) Loop mediated isothermal amplification (LAMP): a new generation of innovative gene amplification technique; perspectives in clinical diagnosis of infectious diseases. Rev Med Virol 18(6):407–421
40. Mori Y, Nagamine K, Tomita N, Notomi T (2001) Detection of loop-mediated isothermal amplification reaction by turbidity derived from magnesium pyrophosphate formation. Biochem Biophys Res Commun 289(1):150–154
41. Thekisoe O, Kuboki N, Nambota A, Fujisaki K, Sugimoto C, Igarashi I, Yasuda J, Inoue N (2007) Species-specific loop-mediated isothermal amplification (LAMP) for diagnosis of trypanosomosis. Acta Trop 102(3):182–189
42. Notomi T, Okayama H, Masubuchi H, Yonekawa T, Watanabe K, Amino N, Hase T (2000) Loop mediated isothermal amplification of DNA. Nucleic Acids Res 28(12), E63
43. Gahlau A, Gothwal A, Chhillar A, Hooda V (2012) Molecular techniques for medical microbiology laboratories: Futuristic approach in diagnostics of infectious diseases. Int J Pharm Bio Sci 3(3):B938–B947
44. Kuboki N, Inoue N, Sakurai T, Di Cello F, Grab D, Suzuki H, Sugimoto C, Igarashi I (2003) Loop Mediated Isothermal Amplification (LAMP) for detection of African Trypanosomes. J Clin Microbiol 41(12):5517–5524
45. Thekisoe O, Rodriguez C, Rivas F, Coronel A, Fukumoto S, Sugimoto C, Kawazu A, Inoue N (2010) Detection of *Trypanosoma cruzi* and *T. rangeli* infections from *Rhodnius pallescens* bugs by Loop Mediated Isothermal Amplification (LAMP). Am J Trop Med Hyg 82(5):855–860

Chapter 13

PCR Techniques in Next-Generation Sequencing

Rashmi S. Goswami

Abstract

With the advent of next-generation sequencing and its prolific use in the clinical realm, it would appear that techniques such as PCR would not be in high demand. This is not the case however, as PCR techniques play an important role in the success of NGS technology. Although NGS has rapidly become an important part of clinical molecular diagnostics, whole genome sequencing is still difficult to implement in a clinical laboratory due to high costs of sequencing, as well as issues surrounding data processing, analysis, and data storage, which can reduce efficiency and increase turnaround times. As a result, targeted sequencing is often used in clinical diagnostics, due to its increased efficiency. PCR techniques play an integral role in targeted NGS sequencing, allowing for the generation of multiple NGS libraries and the sequencing of multiple targeted regions simultaneously. We will outline the methods we employ in PCR amplification of targeted genomic regions for cancer mutation hotspots using the Ampliseq Cancer Hotspot v2 panel (Life Technologies, Carlsbad, CA).

Key words Next-generation sequencing, Multiplex PCR, Hotspot mutation panel, Targeted sequencing, Ion Torrent, AmpliSeq

1 Introduction

The field of next-generation sequencing (NGS) has taken the world by storm over the past 10–15 years, with improvements in both sequencing and computer technology leading to huge advances in the genomics arena. The study of genomics prior to the emergence of NGS technologies involved small, more manageable genomes from microorganisms such as viruses and bacteria, or the examination of single gene disorders, that could be analyzed with the technologies at hand (Sanger sequencing) [1]. Using these techniques the human genome was painstakingly subcloned into artificial chromosomes and sequenced to yield a complete reference genome in 2003 [2]. This reference material is currently the basis for much of the genomic analysis carried out in the present day. For NGS analysis especially, the output from most platforms involves short reads requiring the use of a reference genome for proper alignment which is the foundation for unearthing novel variants [1].

Rajyalakshmi Luthra et al. (eds.), *Clinical Applications of PCR*, Methods in Molecular Biology, vol. 1392,
DOI 10.1007/978-1-4939-3360-0_13,

Despite the ability of NGS technologies to sequence available DNA materials to great depths, sequencing the entirety of large genomes is still clinically prohibitive due to costs of not only sequencing but also data processing, analysis, and storage [3]. In many clinical settings in which whole genome sequencing is not feasible, targeted sequencing techniques are employed to identify variants involved in disease pathogenesis and prognosis. Often specific gene panels or hotspot mutation panels are used, and in institutions in which the clinical sample load is high, these are often more efficient and cost-effective. To fulfill this clinical need, several companies employ target enrichment methods for NGS testing in which genomic regions are selectively captured or amplified from a DNA sample prior to sequencing [3, 4].

There are two main methods by which genomic regions are targeted for sequencing using NGS technologies. One of these techniques is through a process known as hybrid capture, which is often used for regions 1–50 Mb in length [5]. In this technique genomic DNA is fragmented and then hybridized to complementary oligonucleotide probes specifically synthesized and targeted to the genomic regions of interest. These probes are often biotin-labeled, allowing for the probe-genomic DNA hybrid to bind to streptavidin-coated magnetic beads. This allows for specific genomic regions to be isolated from unwanted genomic DNA after application of a magnetic field [6–8], followed by library preparation, sequencing, and data analysis of the regions of interest.

The other technique by which regions of the genome are selected for deep sequencing involves PCR amplification of targeted regions. This is especially useful if the targeted regions of interest are <100 kb in length [5]. This is particularly valuable for samples with degraded DNA, often found in DNA extracted from formalin-fixed paraffin-embedded (FFPE) samples. A number of different PCR techniques can be used for targeted sequencing, including long-range PCR, multiplex PCR, and microdroplet-based PCR. Multiplex PCR for NGS has been employed by commercial companies such as Life Technologies (Carlsbad, CA) with its AmpliSeq panels, whereas microdroplet-based PCR is currently used by RainDance Technologies (Lexington, MA). Long-range PCR can facilitate the sequencing of entire genes or even the mitochondrial genome. Primers are synthesized to generate amplicons that span the targeted region, which often have an average length of 10 kb. Entire genes or genomes can be covered provided the amplicons are designed to overlap one another. Following normalization and pooling, the amplicons are fragmented, and then undergo end-repair followed by ligation of adapters. Libraries for sequencing are created by PCR amplification of these ligated products using the adapters as primers. The final libraries are pooled into one sample tube to undergo sequencing [9].

Multiplex PCR is another technique used to amplify sequences of interest for subsequent sequencing by next-generation techniques. Here, multiple primer pairs are combined in one tube to amplify numerous regions of the genome simultaneously under one set of cycling conditions [10]. In a similar vein, microdroplet-based PCR also entails the use of multiple PCR primers concurrently; however in this technique each PCR reaction uses one pair of primers and occurs within a microdroplet. Briefly, genomic DNA is fragmented to produce 2–4 kb templates, and contained in oil microdroplets together with primers from a primer library using a microfluidic chip. Each droplet is engineered to contain one simplex PCR reaction, and millions of droplets are collected together to undergo thermal cycling under the same cycling conditions. Following amplification, the droplet emulsions are broken, the genomic DNA is removed, and the PCR products are sheared to an average size of 100 bp. Barcode adapters are ligated to the products, the library is amplified to generate enough template, multiple libraries are normalized and pooled followed by sequencing [11].

Multiplex PCR for targeted NGS, especially for mutation panels, is used commonly and commercialized by companies such as Life Technologies. We use the Life Technologies Ion AmpliSeq Cancer Hotspot v2 (Life Technologies, Carlsbad, CA) panel as part of our clinical diagnostic assays at MD Anderson Cancer Center. This panel examines mutational hotspot genomic regions in 50 known oncogenes and tumor suppressor genes using 207 primer pairs. The hotspot regions include areas that collectively contain approximately known 2,800 COSMIC mutations, many of which are clinically actionable with options for targeted therapies. The targeted panel yields 207 amplicons, of an approximate size ranging from 111 to 187 bp (average amplicon length of 154 bp), which are then sequenced using NGS techniques. We outline the methods in preparation of an NGS library using PCR techniques with the Ion AmpliSeq Cancer Hotspot v2 (Life Technologies, Carlsbad, CA) panel here.

2 Materials

2.1 Assay Controls

Controls are necessary to ensure that mutations in hotspot regions are picked up by the NGS assay. Previously tested patient sample(s) or cell line(s) with one, but preferably more mutation(s) in the 50 genes assayed by the AmpliSeq Cancer panel are used as positive controls. Cell lines are the simplest source of material for controls as they are easier to acquire and maintain than material from patients. In addition, the mutations tested from cell lines are more stable and consistent from batch to batch than patient controls. We use cell lines such as H2122 with the following mutations: *TP53* p.C176F and p.Q61L, *KRAS* p.G12C, and *MET* p.N375S, and DLD1 with the following mutations: *KRAS* p.G13D, *FGFR1* p.A266S, *SMO* p.T640A, *FGFR2* p.K367E, *PIK3CA* p.D549N, and *TP53* p.S241F. In addition, a reagent or no template control is run to monitor for false positives secondary to contamination during the PCR process.

2.2 Samples

The advantage to using AmpliSeq panels is that they can be used on very small amounts of DNA making them especially practical for use with DNA from FFPE tissues. We use DNA extracted from macrodissected tumors on 5–10 unstained paraffin sections. Ideally a total of 0.15 μg is preferred; however we have used total DNA amounts lower than 0.01 μg with successful results.

2.2.1 Materials for Library Preparation

1. Ion AmpliSeq™ Library Kit 2.0 containing:
   - 5× Ion Ampliseq HiFi Master Mix.
   - FuPa Reagent.
   - 7.5× Switch Solution.
   - DNA Ligase.
   - Platinum PCR Super Mix High Fidelity.
   - Library Amplification Primer Mix.
   - Low TE.
2. Ion AmpliSeq™ 5× Cancer Hotspot Panel V2.
3. 20× Ion AmpliSeq™ ID Panel.
4. Xpress Barcode Adaptors (1–16 and 17–32 kits).
5. AB 2720 Thermal Cycler.
6. MicroAmp® Optical 96-well Reaction Plates.
7. MicroAmp® Optical 8 cap strip.
8. MicroAmp® Thermal Cycler Plate Cover.
9. MicroAmp® Optical Film Compression Pad.
10. SPRIPlate Super 96-well Magnetic plate.
11. Agencourt® AMPure® XP Beads.
12. Nuclease-free Water.
13. Absolute ethanol.
14. Agilent 2200 TapeStation instrument/Agilent 2100 Bioanalyzer™.
15. Microcentrifuge.
16. Vortex mixer.
17. Pipettors, 1–1000 μL.
18. Multichannel Pipette P10, P50, P200, P1000.

19. Eppendorf 1.5 ml Low-Bind Tubes.
20. Axygen 0.2 ml strip tubes or 0.6 mL tubes.
21. Qubit 0.5 ml clear, thin-wall tubes.

3 Methods

3.1 DNA Dilution

DNA from each sample is diluted to <0.000909 μg/μL (10 ng/11 μL) and 13 μL of each dilution is placed into 0.6 mL tubes or 0.2 mL strip tubes. If the concentration of the sample is <0.000909 μg/μL, then no dilution is needed and 13 μL of undiluted DNA can be used.

3.2 PCR Amplification of Targeted Genomic Regions

1. Prepare enough Master Mix (5× Ion AmpliSeq HiFi Master Mix, 5× Ion AmpliSeq 5× Cancer Hotspot Panel V2, and 20× Ion AmpliSeq Sample ID Panel) for each sample (9 μL total for each sample) as follows:
   - 5× Ion AmpliSeq HIFIMaster Mix 2.0 (4 μL).
   - 5× Ion AmpliSeq Cancer Primer Pool V2 (4 μL).
   - 20× Ion AmpliSeq Sample ID Panel (1 μL).

   Using a multichannel pipette, transfer 9 μL of Master Mix to wells on a chilled 96-well PCR plate (keep chilled by placing on ice or on a cold plate), one well for each sample, including a reagent (no template) control. Transfer 11 μL of DNA from each sample to a well on the 96-well PCR plate. Use 11 μL of water as a reagent (no template) control. Seal the plate with MicroAmp® Optical 8 cap strips. Vortex and quick-spin the plate.
2. Load the plate in the thermal cycler with the following thermal cycling conditions:
   (a) Stage 1 (Hold; enzyme activation): 99 °C for 2 min.
   (b) Stage 2 (Denaturation): 99 °C for 15 s.
   (c) Stage 3 (Annealing): 60 °C for 4 min.
   (d) Repeat **steps 2** and **3** for a total of 20 cycles.
   (e) Stage 4 (Hold): 10 °C (indefinitely; ∞).

   The reaction may remain on the thermal cycler or be stored at 4 °C overnight.

3.3 Digestion of Primer Sequences

1. Place the FuPa reagent on ice. Remove and discard the strip caps from the 96-well plate, and using a multichannel P10 Pipette, add 2 μL of FuPa Reagent to each amplified sample. Pipette up and down to mix. Ensure that no reagent remains in the tips, so that the total volume is 22 μL.

2. Seal the plate with MicroAmp® Optical 8 cap strips. Vortex and quick-spin the plate. Load the plate in the thermal cycler and run under the following thermal cycling conditions:
   (a) 50°C for 10 min.
   (b) 55°C for 10 min.
   (c) 60 °C for 20 min.
   (d) Hold at 10 °C (for up to 1 h).
3. After the program is complete, vortex and quick-spin the plate and proceed immediately to the next step. It is important to note that the FuPa reaction can only be held at 10 °C for up to 1 h and it is critical that the ligation step be followed within this hour. If not, the primer sequences could be over-digested and cause failure of downstream steps.

3.4 Preparation of Barcode Adapters

1. For each barcode, prepare a mix of Nuclease-Free Water, Ion P1 Adapter and Ion Xpress™ Barcode X at a final dilution of 1:4 for each adapter.
2. Add 40 μL of nuclease-free water to each barcode tube, then add 20 μL of the Ion PI Adapter. Vortex to mix each barcode mix thoroughly and quick-spin.

3.5 Ligation of Barcode Adapters to PCR Amplicons

1. Remove and discard the strip caps from the wells containing the FuPa reaction product. For each sample, prepare a total of 6 μL of Master Mix as follows:
   (a) Switch Solution 2.0 (4 μL).
   (b) DNA Ligase 2.0 (2 μL).
2. Add 6 μL of Master Mix to each well using a single-channel P10 pipette. Change tips between each well. Because the master mix is viscous, pipette along the side of the well and ensure that there is no reagent left in the tips.
3. Add 2 μL of Diluted Barcode Adapter Mix to each sample well using a P10 multichannel pipette followed by documentation (one unique barcode per sample). **Note 1.** Seal the reaction plate with a MicroAmp® Optical 8 cap strip. Vortex and quick-spin the plate.
4. Load the plate in the thermal cycler, and run under the following thermal cycling conditions:
   (a) 22°C for 30 min.
   (b) 72°C for 10 min.
   (c) Hold at 10 °C (indefinitely; ∞).
5. After the run is complete, vortex and quick-spin the plate. The reaction may remain on the thermal cycler or be stored at 4 °C overnight.

3.6 Purification of Unamplified Library

1. Pull Agencourt® AMPure® XP Bead Solution 30 min before use, to reach room temperature.
2. Prepare 50 mL of 70 % ethanol in a 50 mL conical tube, by adding 35 mL of 100 % ethanol and 15 mL of Milli-Q water (18 mΩ). Vortex to mix. The ethanol solution must be made fresh with each batch of samples run.
3. Vortex Agencourt® AMPure® beads before use. Using a P1000 Pipette, transfer the volume of beads for the total number of samples being run into a trough. Using a P200 multichannel Pipette, add 45 μL (1.5× sample volume) of Agencourt® AMPure® XP Reagent to each sample, pipette up and down five times to thoroughly mix the bead suspension with the DNA. Change tips between each column of samples. Discard trough with any unused AMPure reagent.
4. Incubate the mixture for 5 min at room temperature.
5. Place the reaction plate onto a 96-well magnetic plate for a minimum of 3 min. Ensure that the solution is clear. Carefully remove and discard the supernatant without disturbing the bead ring. **Note 2**.
6. Pour out about approximately half of the prepared 70 % ethanol into a newly labeled trough. Without removing the plate from the magnet, using a P200 multichannel Pipette, add 150 μL of freshly prepared 70 % ethanol, and incubate for 1 min, then remove and discard the supernatant without disturbing the bead ring. Add an additional 150 μL of freshly prepared 70 % ethanol, incubate for 1 min, then remove and discard the supernatant without disturbing the bead ring. Discard the ethanol trough with any unused ethanol.
7. Keep the plate on the magnet, air-dry the beads at room temperature for 5 min. Ensure that all ethanol droplets are removed from the wells by using a P10 multichannel Pipette to remove any residual ethanol.
8. Remove the plate containing the Ion AmpliSeq™ library from the magnet, and add 50 μL of Low TE to the pellet to disperse the beads. Pipette up and down five times to thoroughly mix.
9. Seal the reaction plate with MicroAmp® Optical 8 cap strip. Vortex and quick-spin the plate. Place the plate back on the magnetic rack for 5 min. Transfer about 40 μL of DNA to a new 96-well plate. This plate consists of stock library and can be stored at −20 °C indefinitely.

Stock DNA can now be used for further amplification using emulsion PCR techniques followed by sequencing on the Ion Torrent Personal Genome Machine (PGM). To proceed to the emulsion PCR step, it is necessary for each sample to have a minimum concentration of 100 pM. Library samples with an average concentration <100 pM should NOT be allowed to proceed to emulsion PCR and should undergo library preparation from the original DNA sample. Libraries with concentrations >100 pM can be pooled together to undergo amplification using emulsion PCR techniques. Depending on which chip will be run on the Ion Torrent PGM, different numbers of samples can be pooled together to run on a chip. At MD Anderson Cancer Center, we pool the following numbers of samples together:

- Twelve patient samples and 1 Control (i.e., cell line) library on a 318Cv2 chip.
- Three patient samples and 1 Control library (i.e., cell line) on a 316 chip.

The concentration of the pooled samples prior to emulsion PCR is calculated to be 20 pM in a total volume of 130 μL and can be stored at 4 °C for up to 1 month.

Overall, PCR techniques continue to be relevant even as the field of genomics continues to advance with newer systems of practice. In fact, it is through PCR techniques that the NGS field has managed to permeate the clinical diagnostic realm and we expect that these practices will rapidly become a routine part of clinical care.

4 Notes

1. In preparing the barcoded adapters it is important when handling them to be especially careful not to cross contaminate. Change gloves frequently and open one tube at a time.
2. During purification of the unamplified library, the supernatant should be carefully removed and discarded without disturbing the bead ring during the wash steps.

References

1. Koboldt DC, Steinberg KM, Larson DE, Wilson RK, Mardis ER (2013) The next-generation sequencing revolution and its impact on genomics. Cell 155:27–38
2. International Human Genome Sequencing Consortium (2004) Finishing the euchromatic sequence of the human genome. Nature 431: 931–945
3. Bodi K, Perera AG, Adams PS, Bintzler D, Dewar K, Grove DS et al (2013) Comparison of commercially available target enrichment methods for next-generation sequencing. J Biomol Tech 24:73–86
4. Mamanova L, Coffey AJ, Scott CE, Kozarewa I, Turner EH, Kumar A et al (2010) Target-enrichment strategies for next-generation sequencing. Nat Methods 7:111–118
5. Claes KB, De Leeneer K (2014) Dealing with pseudogenes in molecular diagnostics in the next-generation sequencing era. Methods Mol Biol 1167:303–315
6. Albert TJ, Molla MN, Muzny DM, Nazareth L, Wheeler D, Song X et al (2007) Direct selection of human genomic loci by microarray hybridization. Nat Methods 4:903–905
7. Gnirke A, Melnikov A, Maguire J, Rogov P, LeProust EM, Brockman W et al (2009) Solution hybrid selection with ultra-long oligonucleotides for massively parallel targeted sequencing. Nat Biotechnol 27:182–189
8. Hodges E, Xuan Z, Balija V, Kramer M, Molla MN, Smith SW et al (2007) Genome-wide in situ exon capture for selective resequencing. Nat Genet 39:1522–1527

9. Ozcelik H, Shi X, Chang MC, Tram E, Vlasschaert M, Di Nicola N et al (2012) Long-range PCR and next-generation sequencing of BRCA1 and BRCA2 in breast cancer. J Mol Diagn 14:467–475
10. De Leeneer K, Hellemans J, De Schrijver J, Baetens M, Poppe B, Van Criekinge W et al (2011) Massive parallel amplicon sequencing of the breast cancer genes BRCA1 and BRCA2: opportunities, challenges, and limitations. Hum Mutat 32:335–344
11. Tewhey R, Warner JB, Nakano M, Libby B, Medkova M, David PH et al (2009) Microdroplet-based PCR enrichment for large-scale targeted sequencing. Nat Biotechnol 27: 1025–1031

Chapter 14

Single-Cell Quantitative PCR: Advances and Potential in Cancer Diagnostics

Chi Young Ok, Rajesh R. Singh, and Alaa A. Salim

Abstract

Tissues are heterogeneous in their components. If cells of interest are a minor population of collected tissue, it would be difficult to obtain genetic or genomic information of the interested cell population with conventional genomic DNA extraction from the collected tissue. Single-cell DNA analysis is important in the analysis of genetics of cell clonality, genetic anticipation, and single-cell DNA polymorphisms. Single-cell PCR using Single Cell Ampligrid/GeXP platform is described in this chapter.

Key words Single-cell analysis, RT-PCR, Quantitative, Cancer, Diagnostics

1 Introduction

There are considerable differences between individual cells even in what looks like a homogenous population of cells. They can exhibit a high degree of variability and express remarkably different responses to the same stimuli. Gene expression measurements are usually performed on large population of cells in a sample containing different cell types; such measurements will result in unknown contribution from these different cell types and will obscure how a transcript in particular is distributed among the cells. In addition, potentially important gene correlation may be missed in such measurements [1]. If we can disintegrate tissues into individual cells that can be sorted and profiled one by one, their responses can be detected with high sensitivity and studied with greater resolution.

Single-cell DNA analysis, in particular genomic DNA, is important and may be informative in the analysis of genetics of cell clonality, genetic anticipation, and single-cell DNA polymorphisms. For most scientists the quantitative transcriptomics in a single cell is much more important, and the quantitative real-timeRT-PCR is the analytical method of choice given its high sensitivity and reproducibility with the ability to detect single

Rajyalakshmi Luthra et al. (eds.), *Clinical Applications of PCR*, Methods in Molecular Biology, vol. 1392,
DOI 10.1007/978-1-4939-3360-0_14, © Springer Science+Business Media New York 2016

molecules, and wide dynamic range. In single-cell biology the absolute abundance of particular mRNAs or microRNAs and their up- or downregulation in a single cell, compared to their neighbor cells, is the goal. The need for quantitativesingle-cell mRNA analysis is evident given the vast cellular heterogeneity of all tissue cells and the inability of conventional RNA methods, like northern blotting, RNase protection assay, or classical block RT-PCR, to distinguish individual cellular contributions to mRNA abundance differences [2].

High-throughput single-cell analysis has a wide range of applications in clinical practice including prenatal diagnostics for assessment of single-gene Mendelian genetic disorders, complex neurological disorders, infectious diseases, and cancer, offering opportunities for single cell-based monitoring of cancer cells and stem cell-based therapies [3].

Clinical implications of single-cell PCR are almost infinitesimal [4]. In cancer genetics, origin of Hodgkin and Reed-Sternberg cell had been controversial for a long time. By using single-cell PCR, it was found to bear clonal immunoglobulin rearrangement, hence proven to be B-cell origin [5]. Single-cell PCR also allowed researchers to compare a morphologically normal lymphocyte and a cell with lobulated nucleus in a patient with adult T-cell leukemia and reveal integration of human T lymphotropic virus type 1 (HTLV-1) in the latter [6]. Furthermore, this technology can also be used for prenatal diagnosis of fetal chromosomal aberrations or genetic disorders, microsatellites, and the genetic identification of fetal cells or other research.

In this chapter we describe single-cell PCR using Single Cell Ampligrid/GeXP platform. In summary, this protocol involves the conversion of specific RNAs within total RNA to cDNA (not oligo dT chemistry or hexamer). The cDNA will be amplified in a multiplex PCR reaction after adding terminal tag sequences. Primers to terminal template sequences are used to reduce primer bias during amplification. There is no plateau in the PCR reaction. There is no saturation at either high-expression or low-expression gene targets. The profile will maintain the expression ratio from wells throughout assay cycle and ratios are locked in within the first five cycles. Then the PCR product will be run on GeXP for analysis.

The advantages of Ampligrid technology include the following: no need for DNA/RNA isolation; low volume of PCR solution, which will result in low cost and low thermal mass; use of glass slide with flat surface which will improve heat conductivity and optical control of the target, respectively; and automated pipetting with high-throughput applications.

2 Materials

2.1 Cell Isolation and Sorting

1. 100 mm cell culture dish with complete media.
2. Trypsin for trypsinization: 3 ml.
3. Media: 7 ml.
4. Serum-free media for resuspension: 1 ml.
5. FACS tube with a 100 μm filter cap.
6. Propidium iodine (1 mg/ml): 10 μl.
7. Hoechst stain (1 mM): 10 μl.
8. Ampligrid slide sorting adapter.
9. Florescence microscope.

2.2 Reverse Transcription PCR

RT master mix calculations (these are based off of a standard GeXP reaction of a total of 20 μl,;depending on how many reactions needed, these volumes can be adjusted):

1. Nucleus-free water: 8 μl.
2. GeXP RT buffer: 4 μl.
3. Reverse Plex primers: 4 μl.
4. Reverse transcriptase: 1 μl.
5. Kana RNA: 3 μl.
6. Ampligrid mineral oil: 5 μl.
7. Amplispeed slide cycler.

2.3 The PCR

1. $MgCl_2$ (25 mM): 4 μl.
2. PCR buffer 5×: 4 μl.
3. Forward multiplex: 2 μl.
4. Thermo-Start Taq: 0.7 μl.
5. cDNA sample (RT): 9.3 μl.

2.4 GeXP

1. Separation gel.
2. Capillary array.
3. SLS loading solution: 37.5 μl.
4. Size standard 0.5 μl.
5. Mineral oil.
6. GeXP software.

3 Methods

1. Grow cells in 100 mm cell culture dish with complete media until about 85–90 % confluent (about ten million cells).
2. Day before sorting, serum starve cells for 24 h.
3. Prepare live cells for sorting:

   Add 3 ml of trypsin for trypsinization, wait for about 3–5 min, add 7 ml media, collect in 15 ml tube, and centrifuge at 250g for 5 min.

   Aspirate media and resuspend with 1 ml of serum-free media.

   Transfer the cells to a FACS tube with a 100 μm filter cap.

   Add 10 μl of propidium iodine (1 mg/ml).

   Add 10 μl of Hoechst stain (1 mM).

   Cover tube with foil and keep on ice until sorting.
4. Go to sorting core with cells.

   Make sure that they have the Ampligrid slide sorting adapter.

   Sort the cells for PI-negative and Hoechst-positive cells so only viable cells will be sorted on to the slide.

   Make sure that they adjust the nozzle to sort one cell per one drop.

   Verify that each reaction site on the slide has a nucleus, or Hoeschst stain nucleus by using a florescence microscope.

   Keep the slides covered and store them in an airtight container at −80 °C until using them for PCR.
5. PCR

   The PCR has two steps, a reverse transcription and a regular PCR step.

   1. Reverse transcription:

      (a) Allow the slide to reach room temperature before using it.

      (b) Prepare master mixes and experimental design which should include control reactions (no-template control-reaction site without a cell sorted, RT minus—standard reaction without including reverse transcriptase, RNA spot—a standard reaction with total RNA spotted onto the slide about 1 μl dried).

      RT master mix calculations (these are based off of a standard GeXP reaction of a total of 20 μl; depending on how many reactions needed, these volumes can be adjusted). The GeXP Start Kit must be purchased, catalog number A21019.

Nucleus-free water	8 μl
GeXP RT buffer	4 μl
Reverse plex primers	4 μl
Reverse transcriptase	1 μl
Kana RNA	3 μl

(c) Each reaction site gets 1 μl of the master mix from above and is covered by 5 μl of Ampligrid mineral oil.

(d) Cycling conditions:

48 °C for 1 min.

37 °C for 5 min.

42 °C for 60 min.

95 °C for 5 min.

4 °C Hold.

2. The PCR:

(a) Make the master mix for the PCR reactions using the following typical GeXP PCR reaction calculations. Each standard GeXP reaction is a total of 10.7 μl, so the following volumes can be adjusted:

PCR buffer (light sensitive)	4 μl
[a]25 mM $MgCl_2$ (not supplied in kit)	4 μl
Forward plex primers	2 μl
[a]Thermo-Start DNA poly (not supplied in kit)	0.7 μl

[a]Highly recommended to use ABgene or Thermo Fisher's Thermo-Start DNA Polymerase, catalog number AB0908.

(b) Add 1 μl of the above master mix to each reaction site on top of the oil. Be sure to check for air bubbles and remove with pipet if necessary.

(c) Cycling conditions:

95 °C for 5 min.

94 °C for 30 s.

55 °C for 30 s.

70 °C for 1 min.

Repeat **steps 2–4** for 44 additional cycles.

4 °C Hold.

Keep the cycler covered to maintain the fluorescence in the PCR buffer.

6. GeXP

   Follow the standard GeXP multiplex protocol.

   Create a slide layout, and include plex, cell type, or other indicators in the names. Select Frag-3 for the method and select the GeXP analysis parameters.

   Install the separation gel, and install the capillary array.

   Perform a gel capillary fill and four cycles of manifold purge.

   Prepare the sample plate by adding 37.5 µl of SLS loading solution and 0.5 µl of size standard for each reaction.

   Prepare the buffer plate: Add separation buffer to each reaction well till it is ¾ full.

   Remove the complete reaction from the slide and put it into the SLS/SS reaction (in a 96-well reaction plate).

   Cover the reaction plate with a seal and vortex the plate or mix by pipetting. Centrifuge (1530g for 1–2 min) the plates and remove the seal. Add one drop of mineral oil to each reaction and centrifuge again (1530g for 1–2 min).

   Load the reaction plate and buffer plate into the GeXP and start run.

4 Notes

In **step 3** of Methods (Prepare live cells for sorting), double-check that the media color will not interfere with the PCR process.

At the end of **step 3** of Methods (Prepare live cells for sorting), please make sure that you have single-cell suspension and healthy cells.

In **step 4** of Methods (Go to sorting core with cells), it is very important to calibrate for one living cell/drop. Also please ensure that there is no dusts on the slides, which can interfere with the PCR. In **step 5** of Methods (PCR), two-step PCR provides more flexibility.

In **step 5.1**(**a**) of Methods, make sure that the slides are not wet. **Step 5.1**(**b**) of Subheading 3 is very important for QC.

In **step 5.1**(**c**) of Methods, use 2 µL pipette to add exact amount of master mix. Be extremely careful not to introduce air bubble while layering the reaction with oil. Make sure that you do not touch the pipette tips to the master mix and change the tip after each step. Do not use a multichannel pipette for these steps.

Step 5.2(**b**) of Methods is a very critical step. Be extremely careful while adding the solution from top and make sure that the solution transfer to the reaction site. Mix well and spin down before cycling.

References

1. Ståhlberg A, Bengtsson M et al (2010) Single-cell gene expression profiling using reverse transcription quantitative real-time PCR. Methods 4:282–288
2. Eberwine J et al (2001) Single-cell molecular biology. Nat Neurosci 4:1155–1156
3. Devonshire AS, Baradez M-O, Morley G, Marshall D, Foy CA et al (2014) Validation of high-throughput single cell analysis methodology. Anal Biochem 452:103–113
4. Hahn S, Zhong XY, Troeger C et al (2000) Current applications of single-cell PCR. Cell Mol Life Sci 57:96–105
5. Vockerodt M, Soares M, Kanzler H et al (1998) Detection of clonal Hodgkin and Reed-Sternberg cells with identical somatically mutated and rearranged VH genes in different biopsies in relapsed Hodgkin's disease. Blood 92:2899–2907
6. Miyagi T, Murakami K, Sawada T et al (1998) A novel single cell PCR assay: detection of human T lymphotropic virus type I DNA in lymphocytes of patients with adult T cell leukemia. Leukemia 12:1645–1650

Chapter 15

Quantitative Real-Time PCR: Recent Advances

Charanjeet Singh and Sinchita Roy-Chowdhuri

Abstract

Quantitative real-time polymerase chain reaction is a technique for simultaneous amplification and product quantification of a target DNA as the process takes place in real time in a "closed-tube" system. Although this technique can provide an absolute quantification of the initial template copy number, quantification relative to a control sample or second sequence is typically adequate. The quantification process employs melting curve analysis and/or fluorescent detection systems and can provide amplification and genotyping in a relatively short time. Here we describe the properties and uses of various fluorescent detection systems used for quantification.

Key words Polymerase chain reaction (PCR), Quantitative PCR, Real-time PCR, q-PCR, Nucleic acid quantification, Intercalation probe, Hybridization probe, Molecular beacon, Minimal residual disease, Fluorescent genotyping

1 Introduction

Quantitative real-time polymerase chain reaction, also called quantitative polymerase chain reaction and real-time polymerase chain reaction, is a technique based on the conventional polymerase chain reaction (PCR) which is used to amplify and simultaneously quantify a target DNA. By convention of the Minimum Information for Publication of Quantitative Real-Time PCR Experiments (MIQE) guidelines [1] and for uniformity, the abbreviation q-PCR is henceforth used in this chapter to designate quantitative real-time PCR. Also, important to note that it is advised not to use the designation RT-PCR for quantitative/real-time PCR to avoid confusion with reverse transcription PCR.

As opposed to the conventional PCR in which quantification, if performed, is based on "end-point" analysis of the amount of the amplicon, q-PCR allows for the product quantification as the process takes place in real time. Conventional PCR with end-point analysis usually requires additional steps such as gel electrophoresis to separate the PCR product, and although quantification can be performed, it is not a simple task. In contrast, q-PCR allows PCR

Rajyalakshmi Luthra et al. (eds.), *Clinical Applications of PCR*, Methods in Molecular Biology, vol. 1392,
DOI 10.1007/978-1-4939-3360-0_15, © Springer Science+Business Media New York 2016

quantification and analysis during amplification and the amplicon is measured in a "closed-tube" system [2] by using various detection systems that employ melting curve analysis (as discussed in a different chapter) and/or fluorescent detection systems (as discussed below). The dynamic nature of q-PCR that can detect, quantify, as well as genotype is what sets it apart from other techniques. An optical thermocycler is used to monitor the fluorescent molecules through excitation and quantification of the fluorescent emission. Although many different designs for fluorophores are possible, including being linked to an oligonucleotide to form a labeled primer or probe, or being directly bound to double-stranded DNA, the common feature is that there must be a change in fluorescence during PCR that would allow detection and quantification of the product in real time.

In an optimized q-PCR reaction, the concentration of initial template usually determines the number of cycles necessary before the fluorescence rises. The initial exponential amplification is not observable as the concentrations are below the limit of detection. This is followed by a growth phase, and finally a plateau phase [3]. The point of maximum acceleration of this growth curve correlates with the concentration of the initial template and the fractional cycle number is inversely proportional to the log of the initial template concentration [3]. The advantage of a q-PCR assay is the excellent sensitivity (limited only by stochastic variation), a large dynamic range with a precision of 5–10 %. Specificity is typically dependent on the PCR quality and detection method. For absolute quantification standards can be included to provide an exact copy number; however, for most practical purposes a relative quantification (mRNA or diploid DNA) usually provides the necessary information. A "housekeeping gene" (e.g., β-actin, cyclophilin, or glyceraldehyde-3-phosphate dehydrogenase) is often used as a biological reference to normalize results between different experiments [4]. The advantage of using a relative quantification approach is that there is no need for generating a standard calibration curve.

Melting curve analysis (Chapter 7) can be used in conjunction with q-PCR as presumptive identification of the target amplified [5]. SYBR Green I dye is included in the PCR reaction and as the annealed products melt at a constant rate and the strands dissociate, the decrease in fluorescence can be monitored. Probe melting analysis is a convenient genotyping method, using either labeled oligonucleotides or unlabeled oligonucleotides (in combination with saturating DNA dyes). High-resolution melting techniques combined with rapid-cycle PCR expand the power of these techniques [5] and amplification and analysis can be performed in a very short time [6].

q-PCR may also be combined with reverse transcription in order to quantify messenger RNA (mRNA) [2] or microRNA (miRNA). The process of reverse transcription polymerase chain reaction (RT-PCR) generates complementary DNA transcripts and provides a qualitative output in terms of gene expression. By combining RT-PCR with q-PCR one can quantify the gene of interest. This process has numerous implications in cancer genetics. The quantification of mRNA can be done as a one-step process where all steps from cDNA synthesis to PCR amplification occur in a single tube or a two-step process where reverse transcriptase reaction and PCR amplification occur in separate tubes. While a one-step process reduces sample handling and therefore sample contamination risk and inter-experimental variations, it also carries a higher risk of degradation of starting mRNA especially when repeated assays are performed from the same sample.

2 Materials

2.1 Generic Reagents

The preparation of a generic "master mix" containing all reagents that are common to the reaction is highly advised to decrease the variations between reactions. This "master mix" that is devoid of the primer, the probes, or the template can be stored at room temperature for a few days or at 4 °C for a few weeks or at −20 °C for months. Once a dye has been added to the master mix, it is recommended to store the mix in the dark. Shown in Table 1 is an example of a generic master mix [7].

Table 1
Example of a generic master mix

5× Generic reagents	Concentration in final reaction
250 mM Tris, pH 8.3	50 μM
2.5 mg/ml BSA (*see* **Note 1**)	500 μg/ml
1 mM of each dNTP (dATP, dCTP, dGTP, and dTTP)	200 μM each
15 mM $MgCl_2$ (*see* **Note 2**)	3 μM
2 U/10 μl heat-stable polymerase (*see* **Note 3**)	0.4 U/10 μl
5× Dye (*see* **Note 4**)	1×

The table shows a generic reaction mixture, which in this example is shown for five reactions. The right-hand panel shows the final concentration of each reagent that is obtained in the final reaction (*see* **Note 5**)

While it is possible to create generic reagents in-house, many laboratories now use prepared reagent kits for real-time PCR which are tailored for specific chemistries or thermal cyclers and can be purchased from a manufacturer.

List of Supplies and Equipment

1. Real-time thermal cycler (such as ABI 7500/7700 or Roche LightCycler).
2. Autoclaved 1.5 ml microcentrifuge tubes.
3. Thin-walled PCR tubes.
4. MicroAmp Optical 96-well reaction plate.
5. Adjustable pipettes, 1–1000 μl range (pre-PCR).
6. Sterile aerosol-filtered pipette tips, 10–1000 μl sizes.
7. Multichannel adjustable pipette (pre- and post-PCR).
8. Vortex mixer.
9. Quick-spin microfuge.
10. PCR hood.
11. Cryorack.
12. Latex gloves.
13. Biohazard bag and container.

2.2 Oligonucleotide Mixture

The oligonucleotide mixture, containing primers and probes, is typically prepared as a 5× mixture, where the primers are at 2.5 μM and probes at 1.0 μM, resulting in final primer and probe concentration of 0.5 μM and 0.2 μM, respectively, in the reaction (see below for primer and probe design) [7]. The oligonucleotide mixture is prepared in 1× TE (10 mM Tris, 0.1 mM ethylenediaminetetraacetic acid [EDTA], pH 8.0) and can be stored at 4 °C for weeks and at −20 °C for months [7] (*see* **Note 6**).

2.3 Template Solution

Template DNA is best diluted from a concentrated stock solution (10^5–10^{10} copies per μl) just prior to use as some templates (particularly diluted PCR products) are not stable for more than 24 h at 4–25 °C [7]. Template (in 1× TE) is usually diluted two- to tenfold in the final PCR solution (*see* **Note 7**).

2.4 Controls

1. *Reagent control/no-template control*: Consists of all PCR reagents except DNA template to ensure no contamination of PCR reaction.
2. *Negative control*: Typically cell line that is negative for the target and used to ensure specificity of the PCR reaction.
3. *Positive control*: Undiluted or diluted cell lines to ensure precision and prevent run-to-run variation in quantification. A sensitivity control may be included to ensure detection of low-level targets.

3 Methods

3.1 q-PCR Primer Design

Designing a primer for q-PCR is similar to that for a conventional PCR (*see* **Note 8**). As with conventional PCR, at least one forward and one reverse primer are used; the primers used may or may not be tagged with a fluorescent marker. A few useful caveats and guidelines [8] are as follows:

1. *Primer length*: The length of most primers should be between 15 and 30 base pairs. A primer which is more than 17 nucleotide bases in length has a great chance of being unique in the human genome [9]. Longer primer lengths are best used in situations when primer is designed against a low-complexity sequence that is not unique. In such cases, longer sequence can reduce nonspecific binding and improve specificity.
2. *Primer melting temperature*: The melting temperature of both forward and reverse primers should be matched to each other [10] and ideally should be within 1 °C of each other. Furthermore, when hybridization/hydrolysis probes are used, the ideal temperature of primers [11] is about 10 °C less than that of the probe used.
3. *Primer concentration*: It is imperative to get the reliable results that the primer concentration is accurate. If a commercially available working stock/kit is not being used, inter-run errors can be reduced by creating a working stock of the primers and pipetting the same volume routinely. The accuracy of the primer concentration in a laboratory-made working stock can be ensured by measurements of spectrophotometric absorbance.
4. *Primer specificity*: Pseudogenes and related bacterial or viral strains, simple sequence repeats, and common repeated sequences should be best avoided. It is also best to avoid primers that have sequences complementary to internal sequences of the intended product.
5. *Non-transcript-specific amplification*: In reactions involving q-PCR of cDNA obtained after RT-PCR, this is best avoided by using a sound RNA isolation technique. Treatment of the RNA sample with DNase is recommended to remove all nonspecific contaminating DNA before performing the reverse transcription step. Additionally, designing primers to span the intron splice sites of the target mRNA can reduce nonspecific amplification of genomic DNA.
6. *Primer cohesion*: It is best to avoid primers that anneal to each other or to other primers [12], particularly at their 3′ ends. This can be prevented by avoiding primers that have more than three bases complementary to each other.

7. *Amplicon length*: Amplification efficiency reduces as the length of amplicon increases. Shorter products amplify with higher efficiency, so it is best to choose a product length less than 500 bp, and preferably less than 200 bp. The suggested amplicon length for optimal PCR efficiency is 50–150 bases.
8. *Template sequence ambiguities*: Basic Local Alignment Search Tool (BLAST) [13] (http://www.ncbi.nlm.nihgov/BLAST/) and Single Nucleotide Polymorphism database (dbSNP) [14] (http://www.ncbi.nlm.nih.gov/SNP/) provided by National Center for Biotechnology Information (NCBI) are free tools that can, respectively, be used to verify the DNA sequence of interest and check for the presence of single-nucleotide polymorphism (SNP) sites in the template. The accuracy of the template sequence is crucial in preventing the assay failure caused due to poor or no binding between primer and template or probe and template (*see* **Note 9**).

3.2 q-PCR Probe Design

The major advancement that has led to widespread use of q-PCR in the clinical and research laboratories is development of novel fluorescent DNA labeling and detection systems. These fluorescent systems enable the detection of the PCR products/amplicons in real time. All q-PCR fluorescent detection system quantifications are based on the principle that the increase in fluorescent signal is directly proportional to the amount of the PCR product generated in the reaction of interest. The two main detection systems used are intercalation based and hybridization based (*see* **Note 10**). While intercalation-based fluorescent dyes bind to the minor groove of double-stranded DNA, the hybridization-based systems involve binding of fluorescent-tagged oligonucleotide probes to a complementary sequence on the gene of interest. The details of different types of probes are discussed further.

1. *Intercalation probes:* The most commonly used intercalation-based probe is SYBR Green-I [15] which has an excitation wavelength of 494 nm and maximum emission wavelength of 521 nm. Another, less commonly used, intercalation probe is an asymmetric cyanine dye called 4-[(3-methyl-6-(benzothiazol-2-yl)-2,3-dihydro-(benzo-1,3-thiazole)-2-methylidene)]-1-methyl-pyridinium iodide (BEBO) [16]. While SYBR green is often detected in channel F1 of the PCR cycler, its alternative BEBO is detected on the FAM channel [16]. Non-molecular probes such as ethidium bromide, also based on the principle of intercalation, may be used; however, its use in clinical and research laboratories has declined in the recent years.

   Both SYBR Green [15] and BEBO work by binding to the minor groove of the double-stranded DNA molecule. At the beginning of the PCR cycle the unbound probe produces a weak fluorescence signal that is interpreted as background signal by the computer software. This background signal can be subtracted from the amplification signal to get the final result. As the PCR cycle progresses, the annealing of the primer to the DNA strand results in binding of the intercalation probe to the double strand. Binding of the probe, to the minor groove of the double-stranded DNA, results in increase in the fluorescence signal [17]. The intensity of the signal increases with the extension step. The amplification fluorescence signal at the end of extension step of the PCR cycle is proportionate to the amount of the amplified DNA. As DNA denaturation occurs in preparation for the next PCR cycle, the intercalation probe molecules get detached resulting in the decrease of fluorescence signal. Amplification signal is measured at the end of each cycle until sufficient/desired PCR product has been accumulated.

SYBR Green or BEBO intercalation probes provide the advantage of sensitive binding to double-stranded DNA, which also makes them relatively easy to use and inexpensive since primer designing and optimization are not required. However, lack of specificity of these probes limits their use. Not uncommonly the PCR reactions also amplify nonspecific products such as homologs and primer dimers. The binding of the primers to themselves forms the "primer dimers," while homologs result when a one or two base pair mismatch happens in the primer-binding site. This small mismatch in a homolog may still allow DNA to be amplified with fidelity by the polymerase. Although these nonspecific products are not related to the target DNA, the intercalation probes also tend to bind to these products. Therefore, the final amplification fluorescence signal is the sum of the target DNA fluorescence and nonspecific product fluorescence. The shape of the amplification curve cannot determine whether the majority of the signal produced is from the target or from the nonspecific product.

However, more recent advances in the PCR cyclers incorporate a melting curve analysis of the product after the final PCR cycle. This is based on the premise that the target DNA and nonspecific products have different sequences and GC content, and therefore their melting curves should be different. Such analysis can add specificity to the use of these probes albeit caution is warranted while differentiating target DNA from its homologs with relatively similar melting curve. In such cases where the use of melting curve analysis is limited by its low resolution, sequencing of the product may be performed. Both sequencing and melting curve analyses add to the procedure and the analysis time. Alternatively, the use of hybridization probes can overcome the issues of nonspecific products/signals and additional analyses.

2. *Hybridization probes:* The hybridization probe [18] based systems work on the principle of fluorescence resonance energy transfer (FRET) [19], i.e., transfer of energy from a donor to an acceptor fluorophore. The donor fluorophore typically emits the light of shorter wavelength that is "quenched" by the acceptor fluorophore when both the fluorophores are in close proximity. However, this transfer or FRET does not occur if the fluorophores are separated and the fluorescence signal emitted by donor fluorophore can be measured [20]. Different designs of hybridization probes show improvement in specificity of the probes and measurement of fluorescence, as discussed below.

   (a) Hydrolysis probes.

   The hydrolysis probes are oligonucleotide sequences that attach downstream to the primer location and utilize the 5′ exonuclease activity of the DNA polymerase enzyme. In the hydrolysis probes, the donor fluorophore is attached to the 5′ end of the oligonucleotide while the acceptor fluorophore is located at the 3′ end. The acceptor at 3′ end not only quenches the fluorescence signal from the donor, but also has an added advantage of blocking the 3′ end of the probe, thereby preventing nonspecific extension of the probe by DNA polymerase.

   During the annealing phase of the cycle the oligonucleotide probe hybridizes to the target DNA at a complementary sequence. The fluorescent signal released from the acceptor fluorophore, both in the reaction mixture and upon annealing, can be detected during the early phase of the PCR and this forms the baseline signal. Subsequently, if the target of interest is present in the reaction mixture, when the thermostable DNA polymerase causes extension of the primer sequence and eventually arrives at the 5′ end of the probe, the donor fluorophore is cleaved by 5′-exonuclease activity of the DNA polymerase. This results in decoupling of the donor and the acceptor fluorophores. The fluorescence from the cleaved donor fluorophore causes the signal to rise above the baseline. The increased fluorescence signal is sometimes also referred to as "Förster-type energy transfer." The intensity of fluorescence amplification signal during the exponential phase of q-PCR is proportional to the number of the donor probes cleaved, which in turn is proportionate to the number of amplicons produced. Thus, the fluorescence signal intensity can be used to quantitate the amount of target DNA in real time, with each cycle.

   The total amount of amplicons and the fluorescence signal intensity increase with each new PCR cycle. If the starting amount of DNA in a sample at the beginning of the first cycle is unknown, mathematical calculations can be used to deduce this amount once the amount of the final DNA product and the total number of PCR cycles are known. This calculation forms the basis of molecular tests used for minimal residual disease (MRD) assessment in clinical settings. Additionally, the fact that hydrolysis probes are located downstream of the primer location and therefore the cleaving of the donor fluorophore occurs only when the target sequence is extended adds to the high specificity needed in MRD testing. A mismatched probe in the setting of a homolog does not bind tightly to its site and the donor fluorophore at the 5′ end, instead of being cleaved, is displaced with its quencher fluorophore. As such, there is no signal amplification from the homolog, although the homolog product may be amplified. Similarly, the final amplification signal is unaffected by primer dimers.

The specificity of hydrolysis probes can be further improved by adding a minor groove binding (MGB) moiety to it—therefore TaqMan-type hydrolysis probes can be either MGB or non-MGB. The use of MGB moiety can increase the melting temperature of the probe by several degrees, which is an important consideration in primer/probe designing.

(b) Dual-hybridization probes.

The dual-hybridization probes further add to the specificity of q-PCR process by the use of two sequence-specific oligonucleotide fluorescent probes instead of one. While the donor fluorophore is located on the 3′ end of one of the probes, the acceptor fluorophore is located at the 5′ end of the second probe. The probes are synthesized such that, upon annealing to the sequence of interest, the 3′ end of the donor fluorophore is in close proximity to the 5′ end of the acceptor fluorophore [21]. The preferred distance between the 3′ end of the first probe and 5′ end of the second probes is between 1 and 5 nucleotides, in order for acceptor fluorophore to quench the energy emitted by the donor fluorophore.

The probes are not in close proximity in the PCR reaction mixture, at the beginning of the reaction. Upon annealing to the sequence of interest downstream from the primer, as the fluorophores are brought together, the low-wavelength light emitted by the donor fluorophore excites the neighboring acceptor fluorophore which in turn emits a fluorescence signal of higher wavelength [21]. This fluorescent signal emitted by the acceptor fluorophore is measured by the instrument. The amount of fluorescence measured is proportional to the amount of DNA generated during the ongoing PCR process. As the primer elongates during the PCR process, the probes either melt off from the template due to rising temperature or are displaced by the DNA polymerase enzyme activity. Both these processes cause the distance between the probes to increase and fall in the fluorescent signal intensity.

(c) Molecular beacons.

Molecular beacons are oligonucleotide probes which utilize two principles [22]. They have a donor fluorophore attached to the 5′ end and an acceptor fluorophore attached to the 3′ end of the oligonucleotide, similar to the hydrolysis probes. However, these probes have a stem-and-loop structure when not bound to the target sequence. The loop has a sequence that is complementary to the target sequence being detected while the stem is formed by two arms. The short, flanking arms have complementary sequences that cause them to anneal to each other. The fluorophores located at the ends of each arm are brought in close proximity by the formation of stem structure [23]. In this configuration, the fluorescence is quenched by energy transfer between the two fluorophores.

The second principle is that the hybrid of loop-target is specific, longer, and more stable than the stem structure formed by the arms. Due to this reason, upon annealing of the loop sequence with the target sequence, the probe changes its conformation from loop-stem to linear structure. The change in conformation causes the arm sequences and therefore the fluorophores to be separated. The acceptor fluorophore in this linear configuration can no longer quench the energy from the donor fluorophore. The fluorescence signal intensity, thus increased, signifies specific target DNA binding. Variations in the sequence of the target DNA by even a single nucleotide make it noncomplementary to the sequence of the loop. Any hybrid between the loop and a noncomplementary target decreases its stability and makes the stem structure of the molecular beacon more stable [23]. Therefore, the conformational change from loop-stem to linear structure does not occur and the mismatched loop-target helix unhybridizes. Their specificity and the length of oligonucleotide with its arms often make the molecular beacons more expensive than hydrolysis or dual-hybridization probes [24].

3.3 Reaction Preparation

A final master mixture can be prepared by adding one volume of the 5× generic mixture, one volume of the 5× oligonucleotide solution, and two volumes of water in a microfuge tube on ice.

Table 2
Example of reagent master mixture

	For one 25 μl reaction (μl)	For n 25 μl reaction (μl)
DEPC-treated water	10	10*n*
5× Generic reagents	5	5*n*
5× Oligonucleotide mixture	5	5*n*
5× Template	5	5*n*

The table shows an example of a reagent master mixture. The middle column shows the pipetted amounts for a single 25 μl reaction. The values in the middle column can be multiplied by the absolute number of reactions in a run, in order to obtain the amount of reagents in that particular run

Table 3
Example of reaction preparation

	For one 25 μl reaction (μl)	For n 25 μl reaction (μl)
DEPC-treated water	2.5	2.5*n*
2× Universal Master Mix Part No: 4324020	12.5	12.5*n*
5× Oligonucleotide mixture	5	5*n*
5× Template	5	5*n*

The table shows an example of a reaction preparation. The middle column shows the pipetted amounts for a single 25 μl reaction, which can be multiplied by the absolute number of reactions, if multiple reactions are being run

Mix the solution and aliquot 20 μl of the master mix for each reaction and then add 5 μl of template solution (Table 2). Remember to include a no-template negative control.

Alternatively a manufacturer-recommended reagent kit may be used. Shown in Table 3 is an example of a reaction preparation using the reagent master mix from Applied Biosystems, Life Technologies.

Vortex the plate/tubes, spin down in a centrifuge briefly, and immediately place in the thermal cycler for PCR.

3.4 Temperature Cycling

Place the sample into the real-time thermal cycler and program the desired temperature conditions for PCR [25]. For genomic DNA, an initial denaturation of 10 s at 95 °C is more than adequate, except when heat-activated polymerases are used, in which case the instructions from the manufacturer should be followed. Typically for each cycle, the denaturation temperature does not need to be held, except in cases where heat-activated polymerases are used. Optimal annealing temperatures depend on the T_m of the primers and melting analysis can be used to determine the product T_m to establish necessary denaturation temperatures. With primer concentrations of 0.5 μM, annealing is rapid [26] and lowering the annealing time improves specificity. Extension times and temperatures depend on the target amplified. A target that is GC rich will extend faster at higher temperatures (around 80 °C), while lower temperatures (around 70 °C) may be required for AT-rich targets. Extension temperatures of 70–74 °C work well for most reactions and are commonly used. Extension times depend on the length of the target. Smaller targets less than 200 bp do not require more than 10 s of extension, while targets up to 500 bp require 15–20 s of extension time, and targets of 1000 bp need up to 30–60 s of extension. In real-time PCR (as in conventional PCR) a long final extension in not required [26]. Before melting analysis, the products should be denatured (95 °C for 0 s) and rapidly cooled (−20 °C/s) to 10 °C below the lowest expected melting transition. The samples can also be stored at 4–25 °C for at least 24 h before analysis on separate instruments.

3.5 Data Analysis

Each dilution should be run in triplicate. The average CT values from each dilution are then plotted vs. the absolute amount of standard present in the sample to generate a standard curve. Comparison of experimental CT values to this standard curve produces an estimate of the amount of target present in the initial sample. For absolute quantification, standards that bracket the potential concentration range are run in parallel [27]. Typically, five to seven standards are separated in concentration by factors of 10. A negative control (previously analyzed cell line that is negative for the translocation or mutation) and a no-template control should also be included in order to identify any background signal, although it is not used in quantitative analysis. A positive control from a previously analyzed cell line or patient sample is included on each run. Amplification plots or "growth curves" are displayed by plotting fluorescence vs. cycle number. The amplification curves should ideally have an exponential, a linear, and a stationary phase [27]. Ensure that the threshold is placed at the exponential phase for the most accurate determination of starting copy number. Also, make sure that the ending baseline setting is at a cycle before the earliest geometric curve growth. A quantitative standard curve is constructed by relating the log of the initial number of templates in each standard to a fractional cycle number derived from each curve [4]. Although a threshold fluorescence level can be used to define these fractional cycle numbers, the second derivative maximum is more precise and depends on the shape of the curve rather than on a specific fluorescence value. Unknown concentrations are determined by interpolation from the standard curves. If standards are run, then the optimal slope of the standard curve is −3.3 with a correlation coefficient (R^2) as close to 1 as possible [4]. Removing obvious outliers from the standard curve will help obtain a slope and R^2 closer to the optimal.

For relative quantification, absolute standards are not necessary. The relative concentration of target between an experimental and a control sample depends on their fractional cycle numbers and the PCR efficiency [28]. For well-optimized PCR reactions, the efficiency is close to 2.0 and the relative concentration depends only on the fractional cycle numbers. When greater accuracy is required, the PCR efficiency can be derived from the standard curve slope of template dilutions. In some experimental designs, the overall amount of nucleic acid (e.g., cDNA) may be difficult to control. In this case, it is common to normalize the test gene against a reference gene, such as a "housekeeping gene" that is assumed not to vary with the experimental conditions.

3.6 Analytical Sensitivity of q-PCR

The analytical sensitivity of q-PCR assay is defined by the minimum number of specific DNA or RNA targets that can be reliably and reproducibly discriminated from nonspecific background. The analytical sensitivity is usually determined by the end-point requirements of a particular test. The analytic sensitivity is often affected by how sensitive the test needs to be for clinical decision making [29]. In a laboratory, the actual analytic sensitivity is dependent on factors such as quantity of template, quality of the template, PCR conditions, and the detection method. Once the required sensitivity has been established, sensitivity standards can be made by serially diluting the target DNA or RNA from positive cells with that from negative cells. Each of these standards is then tested by PCR amplification. The sensitivity of the assay is given by the dilution that yields the faintest detectable signal. Multiple reactions are then performed at this dilution to confirm the reproducibility of the faintest detectable signal. Sensitivity standards (weak positive controls) must be included in each PCR assay to assure that the required level of sensitivity (PCR efficiency) is obtained. The details of this process have been discussed in another chapter.

4 Notes

1. BSA is used with PCR in glass capillary tubes to prevent polymerase denaturation on the glass surface [7]. Siliconized tubes can also be used but are less convenient.
2. Most reactions work well with 3 mM $MgCl_2$, although a concentration range of 1–5 mM is recommended, along with annealing temperatures of 50–65 °C, depending on the T_m of the primers [7]. Most buffers supplied by the manufacturer will result in final $MgCl_2$ concentrations in this range. However, if one is having difficulty generating a detectable PCR product, it might be worthwhile to run a simple $MgCl_2$ titration to determine the optimal $MgCl_2$ concentration; additional $MgCl_2$ can be added if needed.

3. Any native or engineered heat-stable polymerase can be used, although their extension rates may differ. Heat-activated polymerases require time for activation and increase the time required for PCR. Using a hot start DNA polymerase lowers the chance of forming misprimed products or primer dimers. The polymerase typically becomes activated after a 10–15-min, 95 °C incubation step Inappropriate annealing and extension can occur before PCR begins. Hot start techniques delay the activation of the polymerase until high temperatures are reached. There are three common types of hot start methods: physical separation, antibodies, and heat-labile structural inactivation of the polymerase.
4. The first dye used in real-time PCR was ethidium bromide, an intercalating dye that binds between the bases of dsDNA. SYBR Green I is one of the most widely used dyes today. SYBR Green I fluorescence is greater than that of ethidium bromide and it can be viewed in the same channel as fluorescein, a common probe label [7]. Typically, real-time PCR kits provide SYBR Green and polymerase at ideal concentrations for efficient amplification and detection.
5. Preparing a master mixture (devoid of primers, probes, and templates) is important in preventing variations between reactions. The master mix can be stored at −20 °C for a few months. If dye is added to master mix, it should be stored in the dark.
6. Oligonucleotide mixture contains all primers and probes. For a 5× mixture, 2.5 μM primers and 1.0 μM probes result in primer and probe concentrations of 0.5 μM and 0.2 μM, respectively [7].
7. The typical amount of template per 10 μl reaction is approximately 1.6×10^4 copies. Template (in 1× TE) is usually diluted two- to tenfold in the final PCR solution [7].
8. Primer length, primer melting temperature, primer concentration, primer specificity, and amplicon length are important considerations while designing a primer for any reaction.
9. Basic Local Alignment Search Tool (BLAST) and Single Nucleotide Polymorphism database (dbSNP) are recommended for verification of the DNA sequence of interest and preventing failures due to lack of binding between primer and template or probe and template.
10. Intercalation probes are sensitive for double-stranded DNA, easy to use, and inexpensive but lack specificity, and therefore can bind homologs and primer dimers. Melting curve analysis and/or product sequencing may be needed to increase specificity. Hybridization probes utilize fluorescent resonant energy transfer (FRET), and are synthesized by tagging primers with

fluorescent markers. The examples include hydrolysis probes, dual-hybridization probes, and molecular beacons. These are expensive but more sensitive and specific than intercalation probes, making them useful in minimal residual disease testing.

References

1. Bustin SA, Benes V, Garson JA et al (2009) The MIQE guidelines: minimum information for publication of quantitative real-time PCR experiments. Clin Chem 55(4):611–622
2. Nolan T, Hands RE, Bustin SA (2006) Quantification of mRNA using real-time RT-PCR. Nat Protoc 1(3):1559–1582
3. Burns MJ, Nixon GJ, Foy CA et al (2005) Standardisation of data from real-time quantitative PCR methods – evaluation of outliers and comparison of calibration curves. BMC Biotechnol 5:31
4. Huggett J, Dheda K, Bustin S et al (2005) Real-time RT-PCR normalisation; strategies and considerations. Genes Immun 6(4):279–284
5. Ririe KM, Rasmussen RP, Wittwer CT (1997) Product differentiation by analysis of DNA melting curves during the polymerase chain reaction. Anal Biochem 245(2):154–160
6. Reed GH, Wittwer CT (2004) Sensitivity and specificity of single-nucleotide polymorphism scanning by high-resolution melting analysis. Clin Chem 50(10):1748–1754
7. Pryor RJ, Wittwer CT (2006) Real-time polymerase chain reaction and melting curve analysis. Methods Mol Biol 336:19–32
8. Mitsuhashi M (1996) Technical report: part 2. Basic requirements for designing optimal PCR primers. J Clin Lab Anal 10(5):285–293
9. Haas SA, Hild M, Wright AP et al (2003) Genome-scale design of PCR primers and long oligomers for DNA microarrays. Nucleic Acids Res 31(19):5576–5581
10. Sugimoto N, Nakano S, Yoneyama M et al (1996) Improved thermodynamic parameters and helix initiation factor to predict stability of DNA duplexes. Nucleic Acids Res 24(22): 4501–4505
11. Rychlik W, Spencer WJ, Rhoads RE (1990) Optimization of the annealing temperature for DNA amplification in vitro. Nucleic Acids Res 18(21):6409–6412
12. Vallone PM, Butler JM (2004) AutoDimer: a screening tool for primer-dimer and hairpin structures. Biotechniques 37(2):226–231
13. Mount DW (2007) Using the basic local alignment search tool (BLAST). CSH Protocol. pdb top17
14. Sherry ST, Ward MH, Kholodov M et al (2001) dbSNP: the NCBI database of genetic variation. Nucleic Acids Res 29(1):308–311
15. Giglio S, Monis PT, Saint CP (2003) Demonstration of preferential binding of SYBR Green I to specific DNA fragments in real-time multiplex PCR. Nucleic Acids Res 31(22): e136
16. Bengtsson M, Karlsson HJ, Westman G et al (2003) A new minor groove binding asymmetric cyanine reporter dye for real-time PCR. Nucleic Acids Res 31(8):e45
17. Morrison TB, Weis JJ, Wittwer CT (1998) Quantification of low-copy transcripts by continuous SYBR Green I monitoring during amplification. Biotechniques 24(6):954–958
18. Ram S, Singh RL, Shanker R (2008) In silico comparison of real-time PCR probes for detection of pathogens. In Silico Biol 8(3–4): 251–259
19. Simon A, Labalette P, Ordinaire I et al (2004) Use of fluorescence resonance energy transfer hybridization probes to evaluate quantitative real-time PCR for diagnosis of ocular toxoplasmosis. J Clin Microbiol 42(8):3681–3685
20. Johansson MK (2006) Choosing reporter-quencher pairs for efficient quenching through formation of intramolecular dimers. Methods Mol Biol 335:17–29
21. Wittwer CT, Herrmann MG, Moss AA et al (1997) Continuous fluorescence monitoring of rapid cycle DNA amplification. Biotechniques 22(1):130–131
22. Kostrikis LG, Tyagi S, Mhlanga MM et al (1998) Spectral genotyping of human alleles. Science 279(5354):1228–1229
23. Tyagi S, Kramer FR (1996) Molecular beacons: probes that fluoresce upon hybridization. Nat Biotechnol 14(3):303–308
24. Fang X, Li JJ, Perlette J, Tan W et al (2002) Molecular beacons: novel fluorescent probes. Anal Chem 72(23):747A–753A

25. Lyon E, Wittwer CT (2009) LightCycler technology in molecular diagnostics. J Mol Diagn 11(2):93–101
26. Wittwer CT, Ririe KM, Andrew RV et al (1997) The LightCycler: a microvolume multisample fluorimeter with rapid temperature control. Biotechniques 22(1):176–181
27. Vu HL, Troubetzkoy S, Nguyen HH et al (2000) A method for quantification of absolute amounts of nucleic acids by (RT)-PCR and a new mathematical model for data analysis. Nucleic Acids Res 28(7):E18
28. Regier N, Frey B (2010) Experimental comparison of relative RT-qPCR quantification approaches for gene expression studies in poplar. BMC Mol Biol 11:57
29. Bustin SA, Mueller R (2005) Real-time reverse transcription PCR (qRT-PCR) and its potential use in clinical diagnosis. Clin Sci 109(4): 365–379

Chapter 16

PCR Techniques in Characterizing DNA Methylation

Khalida Wani and Kenneth D. Aldape

Abstract

DNA methylation was the first epigenetic mark to be discovered, involving the addition of a methyl group to the 5′ position of cytosine by DNA methyltransferases, and can be inherited through cell division. DNA methylation plays an important role in normal human development and is associated with the regulation of gene expression, tumorigenesis, and other genetic and epigenetic diseases. Differential methylation is now known to play a central role in the development and outcome of most if not all human malignancies.

Bisulfite conversion is a commonly used approach for gene-specific DNA methylation analysis. Treatment of DNA with bisulfite converts cytosine to uracil while leaving 5-methylcytosine intact, allowing for single-nucleotide resolution information about the methylated areas of DNA. PCR-based methods are routinely used to study DNA methylation on a gene-specific basis, after bisulfite treatment. Variations of this method include bisulfite sequencing, methylation-specific PCR, real-time PCR-based MethyLight, and methylation-sensitive high-resolution melting PCR. Several whole-epigenome profiling technologies such as MethylC-seq reduced representation bisulfite sequencing (RRBS) and the Infinium Human methylation 450 K bead chip are now available allowing for the identification of epigenetic drivers of disease processes as well as biomarkers that could potentially be integrated into clinical practice.

Key words DNA methylation, CpG islands, Bisulfite conversion, Methylation-specific PCR (MSP), Real-time MSP, MethyLight, Whole-genome methylation profiling, Bisulfite sequencing, Methylation sensitive-high-resolution melting (MS-HRM)

1 Introduction

DNA methylation is an important epigenetic modification that is increasingly recognized as one of the major mechanisms of gene regulation in various biological processes including genomic imprinting, transposable element silencing, stem cell differentiation, and embryonic development (Robertson [1]). DNA methylation patterns have also been shown to undergo changes in cancer and several other diseases. DNA methylation in mammalian cells occurs at the 5′-position of cytosine within the CpG dinucleotide. In general, the CpG dinucleotide can be found in small genomic regions of about one kilo base, called CPG ISLANDS. Although CpG islands account for only about 1 % of the genome, they

Rajyalakshmi Luthra et al. (eds.), *Clinical Applications of PCR*, Methods in Molecular Biology, vol. 1392, DOI 10.1007/978-1-4939-3360-0_16,

contain over 15 % of the total genomic CpG sites. There are about 45,000 CpG islands, most of which reside within or near the promoters or first exons of genes and are unmethylated in normal cells [2]. Much attention in the methylation field has focused on CpG islands, primarily because of the propensity of such sequences to become aberrantly hypermethylated in tumors, resulting in the transcriptional silencing of the associated gene [3–5].

The first generation of methylation detection assays employed the digestion of genomic DNA with a methylation-sensitive restriction enzyme followed by either Southern blot analysis or PCR [6]. Although these techniques are relatively straightforward, problems such as the limited availability of informative restriction sites, the occurrence of false-positive results due to incomplete digestion, and the requirement of large amounts of high-molecular-weight DNA have restricted their use. Many of these methods also employ restriction enzyme digestion, radiolabeled dNTPs, or hybridization probes. These labor-intensive steps limit the use of these methods for high-throughput analyses. A second generation of techniques resulted from the demonstration that treatment of genomic DNA with sodium bisulfite followed by alkaline treatment converts unmethylated cytosines to uracil while leaving methylated cytosine residues intact [7]. The benefit of sodium bisulfite-based assays is that they require very small amounts of DNA and consequently are compatible with DNA obtained from microdissected paraffin-embedded tissue samples [8].

Among the various methods available for DNA methylation analysis, PCR-based analysis of bisulfite-converted DNA is the most commonly used. Different variations of the PCR-based method include methylation-specific PCR (MSP) followed by gel electrophoresis, real-time PCR-based MethyLight, methylation-sensitive high-resolution melting (MS-HRM), and bisulfite sequencing (BSP) [9, 10] (Fig. 1).

Currently, massive parallel sequencing and microarray-based platforms have also been developed for methylome profiling. These include methods that are sodium bisulfite based including MethylC-seq, reduced representation bisulfite sequencing (RRBS), and the Infinium Human methylation 450 K bead chip (Illumina Inc. CA, USA). Other available methods such as methylated DNA immunoprecipitation sequencing (MeDIP-seq), methylated DNA capture by affinity purification (MethylCap-seq), and methylated DNA-binding domain sequencing (MBD-seq) methylated DNA capture by affinity purification (MethylCap-seq) and methylated DNA-binding domain sequencing (MBD-seq) rely on capture of methylated DNA by a monoclonal antibody or by the recombinant methyl-binding domains of *MECP2* or *MBD2*, respectively [11–14].

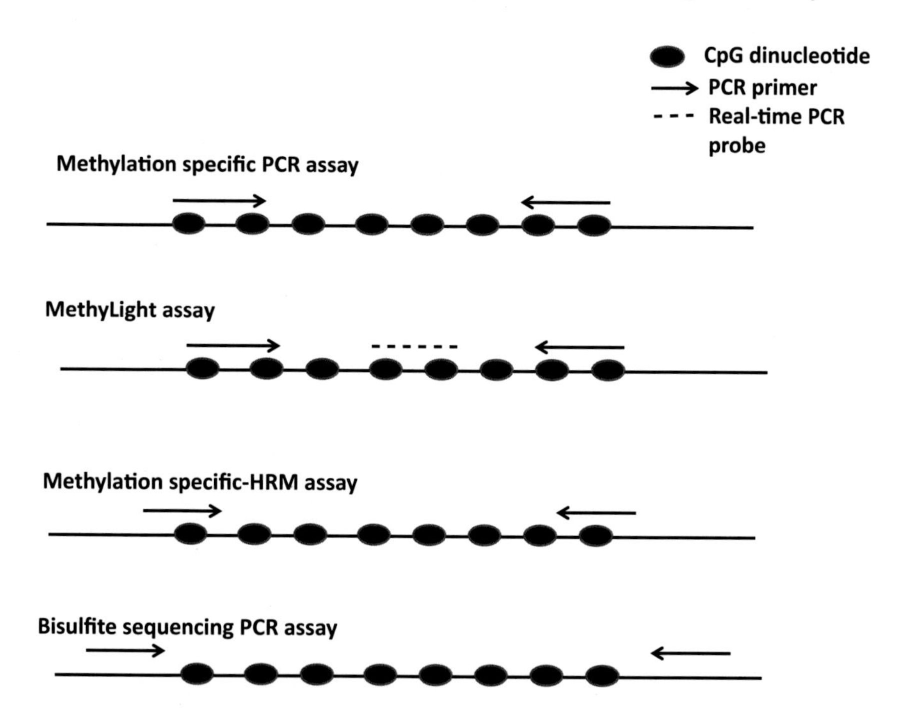

Fig. 1 Schematic representation of the commonly used PCR-based methods for DNA methylation analysis. Methylation-specific PCR, MethyLight, and methylation-sensitive high-resolution melting analysis are methylation-dependant methods which use primers that include CpG sites within their sequence whereas bisulfite sequencing in independent of methylation status since the primers do not cover any CpG sites

2 Materials

2.1 Methylation-Specific PCR (MSP), MethyLight, Methylation-Sensitive High-Resolution Melting, Bisulfite Sequencing

1. Master Pure™ Complete DNA and RNA Purification Kit (Illumina Inc., San Diego, CA).
2. EZ DNA Methylation-Gold™ Kit (Zymo Research, Irvine, CA).
3. AmpliTaq Gold® 360 Master Mix (Life Technologies, Grand Island, NY).
4. TaqMan® Universal PCR Master Mix, No AmpErase® UNG (Life Technologies, Grand Island, NY).
5. Human methylated and non-methylated DNA set (Zymo Research, Irvine, CA).
6. Primers.
7. Fluorescent probes.
8. Isopropanol.
9. Ethanol.

2.2 Equipment

1. Heating block.
2. Microcentrifuge.
3. Thermal cycler.
4. Agarose gel electrophoresis apparatus.
5. Gel documentation system.
6. Applied Biosystems 7500 real-time PCR system.

3 Methods

3.1 Methylation-Specific PCR

First reported in 1996 by Herman et al., methylation-specific PCR [15] has become a widely used technique for analyzing the methylation of CpG sites. Methylation-specific PCR is a PCR-based method for the analysis of methylation patterns in CpG islands. It exploits the sequence variants that arise at a particular locus due to bisulfite conversion using primers designed to anneal with bisulfite-converted DNA. The sequence differences resulting from various DNA methylation patterns can then be revealed in two principally different ways. Either the discrimination is made at the PCR amplification step by the use of primers that anneal specifically with either the converted methylated or converted unmethylated sequence, or the discrimination is left until after the PCR reaction by the use of PCR primers that do not themselves cover any CpG dinucleotides sites in the original genomic DNA.

3.1.1 DNA Extraction: Master Pure™ Complete DNA and RNA Purification Kit

1. Dilute 2 μl of Proteinase K into 300 μl of tissue and cell lysis solution for each sample.
2. Add 300 μl of this lysis buffer to each sample tube and mix thoroughly.
3. Incubate at 55 °C overnight.
4. Place the samples on ice for 3–5 min.
5. Add 175 μl of MPC Protein Precipitation Reagent to lysed sample and vortex mix vigorously for 10 s.
6. Pellet the debris by centrifugation at 4 °C for 10 min at ≥10,000 × *g* in a microcentrifuge.
7. Transfer the supernatant to a clean microcentrifuge tube and discard the pellet.
8. Add 500 μl of isopropanol to the recovered supernatant. Invert the tube several times in order to precipitate the DNA.
9. Pellet the DNA by centrifugation at 4 °C for 10 min in a microcentrifuge.
10. Carefully pour off the isopropanol without dislodging the DNA pellet.

11. Rinse the pellet with 70 % ethanol twice.
12. Remove all of the residual ethanol. Air-dry the DNA pellet for 15 min.
13. Resuspend the DNA pellet in 35 μl of TE buffer. The amount of buffer used for resusupension will depend upon the amount of starting material used for the isolation.
14. Quantify the DNA.

3.1.2 Bisulfite Conversion Protocol: Zymo Research's EZ DNA Methylation-Gold Kit

1. In a PCR tube, add 500 ng of the DNA sample. If the volume of the DNA sample is less than 20 μl, make up the difference with sterile water.

   Note: Each time a sample is run, include a positive and negative control in the bisulfite conversion step. A positive control is a fully methylated commercially available DNA while a negative control is a fully unmethylated commercial DNA sample.
2. Add 130 μl of the prepared CT conversion reagent to the tube. Mix the sample by flicking the tube or pipetting the sample up and down, and then centrifuge the liquid to the bottom of the tube.
3. Place the sample tube in a thermal cycler and perform the following steps:
   (a) 98 °C for 10 min.
   (b) 64 °C for 2.5 h.
   (c) 4 °C storage up to 20 h.

   Note: The following steps must be carried out within 24 h after the cycling above is completed.
4. Place a Zymo-Spin™ IC Column into a provided collection tube, and add 600 μl of M-binding buffer. Let stand for 5 min.
5. Load the sample (from **step 3**) into the Zymo-Spin™ IC Column containing the M-binding buffer. Close the cap and mix by vortexing briefly for 1–2 s.
6. Centrifuge at full speed (>10,000 × *g*) for 30 s. Discard the flow-through.
7. Add 100 μl of M-wash buffer to the column. Centrifuge at full speed for 30 s.
8. Add 200 μl of M-desulfonation buffer to the column and let stand at room temperature (20–30 °C) for 15–20 min. After the incubation, centrifuge at full speed for 30 s. Discard the flow-through.
9. Add 200 μl of M-wash buffer to the column. Centrifuge at full speed for 30 s. Add another 200 μl of M-wash buffer and centrifuge for an additional 30 s.

10. Place the column into a 1.5 ml microcentrifuge tube. Add 60 μl of M-elution buffer directly to the column matrix. Centrifuge for 30 s at full speed to elute the DNA. The DNA is ready for immediate analysis or can be stored at −20 °C for later use. For long-term storage, store at −70 °C.

3.1.3 Primers and Probes

For methylation-specific PCR, two primer sets are designed to anneal to a region containing CpG motifs. The methylated primer set assumes that the CpGs are fully methylated while the unmethylated primer set anneals to genomic DNA that is not methylated in the same primer-binding site. It is important to test the primer sets with a control genomic DNA of known methylation status along with the genomic DNA of unknown methylation status. The properly designed methylated primer set will only amplify the control methylated genomic DNA, and not the unmethylated genomic DNA, whereas unmethylated primer set will amplify only the unmethylated genomic DNA. A reference gene must be included in the assay for normalization.

Methyl Primer Express® Software is a free online primer design tool specifically for methylation studies which assists in designing primers in both methylated and unmethylated bisulfite-modified DNA. Users simply cut and paste in the selected genomic sequence, the software identifies CpG islands, then performs an in silico bisulfite conversion (Cs are converted to Ts), and aids in the selection of primers. Methyl Primer Express software is available for free download at the Life Technologies website.

Recommendations for CpG Island Prediction and Primer Selection

Selecting Region of Interest for MSP

Concerning the predicted CpG islands, several rules govern the choice of appropriate regions as targets for primer design:

1. If more than one island is predicted, any of them can be the target region for amplification.
2. If a CpG island size is smaller than the minimal product size, the primer pair should span the whole island.
3. If a CpG island size is greater than the maximal product size, the primer pair should be within the island.
4. If a CpG island size is between the minimal and maximal product size, primer pair should involve at least two-thirds of the island.

Primer Design

1. Each primer should contain at least three CpG sites in its sequence, and one of the CpG sites should be at the 3′ end of the sequence to maximally discriminate between methylated DNA and unmethylated DNA.
2. CpG sites should be at the 3′ end of the sequence to maximally discriminate between methylated DNA and unmethylated DNA.

3. Each primer should have a maximum number of non-CpG Cs in its sequence to amplify only the bisulfite-modified DNA.
4. The primer pair for the methylated DNA (M pair) and the primer pair for the unmethylated DNA (U pair) should contain the same CpG sites within their sequences. However, they may vary in length or at their start point for matching of temperature value between two sets of primers.
5. Two pairs of primers having similar annealing temperature are preferred and annealing temperatures equal or higher than 55 °C are preferred.

MSP-Agarose Gel-Based Detection

After primer design and selection, MSP is performed with general PCR conditions. For example, PCR conditions consist of 5 min at 94 °C for initial denaturation, followed by 40 cycles of 94 °C (30 s), annealing temperature of primers (30 s), and 72 °C (60 s), and a final elongation of 7 min at 72 °C. Finally, amplified products are loaded onto a 2 % agarose gel and electrophoresed for 20 min at 100 V. An amplification product of the correct molecular weight on an electrophoresis gel can be interpreted as methylated or unmethylated, depending on the specific primers used. The presence of amplification products using both sets of primers indicates a sample with both methylated and unmethylated DNA in the region of interest.

3.2 MSP-MethyLight

MethyLight is a sensitive, high-throughput methylation assay that uses real-time methylation-specific PCR technology to measure DNA methylation [16, 17]. MethyLight does not require post-PCR manipulations of DNA such as gel electrophoresis, since the analysis is performed at the PCR level. Several variations of MethyLight have been proposed to address different biological questions, such as the amount of methylated versus unmethylated alleles or methylation status at the CpG dinucleotide level. The most commonly used MethyLight methodology uses two primers and a TaqMan probe designed to bind the methylated allele specifically and requires a reference gene for normalization [18].

3.2.1 Real-Time PCR

1. Add 10 μl of 2× TaqMan® Universal PCR master mix per reaction in a 96-well plate.
2. Add 6 μl of 3 mM primer probe mix.
3. Add 2 μl of bisulfite-converted DNA. Make up the final volume to 20 μl with sterile distilled water.
4. Seal the plate using the Micro Amp optical adhesive film.
5. Centrifuge plate at 1000 × *g* for 1 min.
6. Run the real-time PCR on an Applied Biosystems 7500 real-time PCR system.

7. Cycling conditions:
   (a) 90 °C for 10 min—Amplitaq Gold DNA polymerase activation step.
   (b) 90 °C for 15 s—Denaturation step.
   (c) 60 °C for 1 min—Annealing/extension step.

   Repeat **steps b** and **c** for a total of 40–45 cycles.
8. Semiquantitative DNA methylation levels are determined by calculating delta Ct values for the gene of interest to a reference gene such as COL2A1 or ACTB.

3.3 Methylation-Sensitive High-Resolution Melting

High-resolution melting technology is based on the comparison of the melting profiles of sequences that differ in base composition and has been shown to have a sensitivity superior to direct sequencing [19]. Each double-stranded DNA molecule is characterized by a melting temperature or, for longer molecules, a melting profile. The melting temperature is defined as the specific temperature at which the DNA helix dissociates into two single strands and is sequence dependent. The melting temperature of a PCR product can be investigated by subjecting it to an increasing temperature gradient in the presence of a DNA intercalating dye, which emits fluorescence when bound (intercalated) to double-stranded DNA. The fluorescent dye will bind to double-stranded DNA emitting high levels of fluorescence until the temperature reaches the melting temperature of the PCR product. At the melting temperature, the PCR product dissociates into two single strands and the dye can no longer bind and fluoresce, and a sharp drop in the fluorescence is observed. The changes of fluorescence levels across a denaturing gradient describe an amplicon's melting profile. The PCR product originating from the methylated allele will have different sequence composition (GC content) from the PCR product derived from unmethylated variant of the same locus. As a consequence, both products have distinct melting temperature and melting profiles. Hence, to investigate the methylation status of an unknown locus, the melting profile of the PCR product derived from that locus has to be compared with the profiles of methylated and unmethylated controls [20]. The melting analyses do not allow detailed information about the methylation of single cytosines within the sequence of interest to be obtained. These can be assessed only by DNA sequencing technologies. Nevertheless, MS-HRM analysis can distinguish fully and partially methylated samples. Therefore, MS-HRM can identify those samples for which further sequence information would be of interest by selecting them from the samples that are clearly unmethylated and thus do not require sequencing.

3.4 Bisulfite Sequencing

Bisulfite sequencing PCR (BSP) was the first technique described for analyzing DNA methylation status using PCR [21]. The technique consists of PCR amplifying a bisulfite-converted DNA region of interest, followed by Sanger sequencing of the product either directly or after cloning into a suitable vector. Methylation-independent PCR primers should be designed to allow the amplification of bisulfite-converted DNA regardless of methylation status. Primers should also not bind regions containing CpG dinucleotides and should flank a sequence of converted DNA containing as many thymines originating from the conversion of non-CpG cytosines as possible. A qualitative analysis of bisulfite sequencing results can be performed if a clear single peak is present for each CpG cytosine position. In that case, a thymine peak would be interpreted as a non-methylated CpG, and a cytosine peak would represent a methylated CpG.

References

1. Robertson KD (2005) DNA methylation and human disease. Nat Rev Genet 6:597–610
2. Antequera F, Bird A (1993) Number of CpG islands and genes in human and mouse. Proc Natl Acad Sci U S A 90:11995–11999
3. Issa JP (2007) DNA methylation as a therapeutic target in cancer. Clin Cancer Res 13: 1634–1637
4. Portela A, Esteller M (2010) Epigenetic modifications and human disease. Nat Biotechnol 28:1057–1068
5. Heyn H, Esteller M (2012) DNA methylation profiling in the clinic: applications and challenges. Nat Rev Genet 13:679–692
6. Singer-Sam J, LeBon JM, Tanguay RL, Riggs AD (1990) A quantitative HpaII-PCR assay to measure methylation of DNA from a small number of cells. Nucleic Acids Res 18:687
7. Frommer M, McDonald LE, Millar DS, Collis CM, Watt F, Grigg GW, Molloy PL, Paul CL (1992) A genomic sequencing protocol that yields a positive display of 5-methylcytosine residues in individual DNA strands. Proc Natl Acad Sci U S A 89:1827–1831
8. Zhu J, Yao X (2007) Use of DNA methylation for cancer detection and molecular classification. J Biochem Mol Biol 40:135–141
9. Hernandez HG, Tse MY, Pang SC, Arboleda H, Forero DA (2013) Optimizing methodologies for PCR-based DNA methylation analysis. Biotechniques 55:181–197
10. Shanmuganathan R, Basheer NB, Amirthalingam L, Muthukumar H, Kaliaperumal R, Shanmugam K (2013) Conventional and nanotechniques for DNA methylation profiling. J Mol Diagn 15:17–26
11. How Kit A, Nielsen HM, Tost J (2012) DNA methylation based biomarkers: practical considerations and applications. Biochimie 94:2314–2337
12. Bock C, Tomazou EM, Brinkman AB, Muller F, Simmer F, Gu H, Jager N, Gnirke A, Stunnenberg HG, Meissner A (2010) Quantitative comparison of genome-wide DNA methylation mapping technologies. Nat Biotechnol 28:1106–1114
13. Harris RA, Wang T, Coarfa C, Nagarajan RP, Hong C, Downey SL, Johnson BE, Fouse SD, Delaney A, Zhao Y et al (2010) Comparison of sequencing-based methods to profile DNA methylation and identification of monoallelic epigenetic modifications. Nat Biotechnol 28:1097–1105
14. Mensaert K, Denil S, Trooskens G, Van Criekinge W, Thas O, De Meyer T (2014) Next-generation technologies and data analytical approaches for epigenomics. Environ Mol Mutagen 55:155–170
15. Herman JG, Graff JR, Myohanen S, Nelkin BD, Baylin SB (1996) Methylation-specific PCR: a novel PCR assay for methylation status of CpG islands. Proc Natl Acad Sci U S A 93:9821–9826
16. Eads CA, Danenberg KD, Kawakami K, Saltz LB, Blake C, Shibata D, Danenberg PV, Laird PW (2000) MethyLight: a high-throughput assay to measure DNA methylation. Nucleic Acids Res 28:E32
17. Campan M, Weisenberger DJ, Trinh B, Laird PW (2009) MethyLight. Methods Mol Biol 507:325–337
18. Trinh BN, Long TI, Laird PW (2001) DNA methylation analysis by MethyLight technology. Methods 25:456–462

19. Wittwer CT, Reed GH, Gundry CN, Vandersteen JG, Pryor RJ (2003) High-resolution genotyping by amplicon melting analysis using LCGreen. Clin Chem 49:853–860
20. Wojdacz TK, Dobrovic A (2007) Methylation-sensitive high resolution melting (MS-HRM): a new approach for sensitive and high-throughput assessment of methylation. Nucleic Acids Res 35:e41
21. Clark SJ, Harrison J, Paul CL, Frommer M (1994) High sensitivity mapping of methylated cytosines. Nucleic Acids Res 22:2990–2997

Index

Rajyalakshmi Luthra et al. (eds.), *Clinical Applications of PCR*, Methods in Molecular Biology, vol. 1392,
DOI 10.1007/978-1-4939-3360-0, © Springer Science+Business Media New York 2016

D

E

F

G

H

N

O

P

Q

Printed in the United States
By Bookmasters